Alfa Romeo

ALFETTA GTV6 1980-1986

Compiled by
R M Clarke

ISBN 9781855202160

BROOKLANDS BOOKS LTD.
P.O. BOX 904, AMERSHAM,
BUCKS, HP6 9JA. UK
sales@brooklands-books.com

AR80 www.brooklandsbooks.com 8W4/2894

ROAD TEST SERIES

Abarth Road Test Portfolio 1950-1971
AC Cars 1904-2011
AC Ace Aceca Greyhound Road Test Portfolio 1953-1962
AC & Cobra 1962-2011
Alfa Romeo A Brooklands Portfolio 1920-1933
Alfa Romeo A Brooklands Portfolio 1934-1940
Alfa Romeo A Brooklands Portfolio 1920-1940
Alfa Romeo Giulietta Gold Portfolio 1954-1965
Alfa Romeo Giulia Berlina Lim. Edit. Extra 1962-1976
Alfa Romeo Giulia Coupes Lim. Edit. Ultra 1973-1976
Alfa Romeo Alfasud 1972-1984
Alfa Romeo Alfetta Gold Portfolio 1972-1987
Alfa Romeo Alfetta GTV6 1980-1986
Alfa Romeo Spider Ultimate Portfolio 1966-1994
Alfa Romeo Spider & GTV Perf. Port. 1995-2005
Allard Limited Edition Ultra
Alpine Renault Ultimate Portfolio 1958-1995
Alvis Gold Portfolio 1919-1967
AMC Rambler Limited Edition Extra 1956-1969
AMX & Javelin Gold Portfolio 1968-1974
Armstrong Siddeley Gold Portfolio 1945-1960
Aston Martin Gold Portfolio 1921-1947
Aston Martin Ultimate Portfolio 1948-1968
Aston Martin Ultimate Portfolio 1968-1980
Aston Martin Ultimate Portfolio 1981-1993
Aston Martin Ultimate Portfolio 1994-2006
Auburn Cord Duesenberg - A Brooklands Portfolio
Audi Quattro Gold Portfolio 1980-1991
Audi TT Performance Portfolio 1998-2006
Austin Seven & Ten Road Test Portfolio
Austin-Healey 100 & 100/6 Gold Port. 1952-1959
Austin-Healey 3000 Road Test Portfolio
Austin-Healey Sprite Bug-Eye Mk 1 Road Test Portfolio
Avanti Limited Edition Extra
Bentley & Rolls-Royce Portfolio 1990-2002
Berkeley Sportscars Limited Edition
BMW 700 Limited Edition 1959-1965
BMW 2002 Ultimate Portfolio 1968-1976
BMW 6 Cylinder Coupes & Saloons Gold P. 1969-1976
BMW 320, 323, 325 (6 cyl.) Gold Port. 1977-1990
BMW 3 Series Gold Portfolio 1991-1997
BMW M3 Ultimate Portfolio 1986-2006
BMW M5 Gold Portfolio 1980-2003
BMW 5 Series Gold Portfolio 1988-1995
BMW 6 Series Ultimate Portfolio 1976-1989
BMW 7 Series Performance Portfolio 1977-1986
BMW 7 Series Performance Portfolio 1986-1993
BMW 8 Series Performance Portfolio
BMW X5 Limited Edition Extra 1999-2006
BMW Alpina Performance Portfolio 1967-1987
BMW Alpina Performance Portfolio 1988-1998
BMW Z3, M Coupe & M Roadster Gold Port. 1996-02
Bond Cars 1949-1974 Road Test Portfolio
Borgward Isabella Limited Edition
Bricklin Golf Portfolio 1974-1975
Bristol Cars Portfolio
Bugatti Type 10 to Type 40 Road Test Portfolio
Bugatti Type 41 to Type 55 Road Test Portfolio
Bugatti Type 57 to Type 251 Road Test Portfolio
Bugatti Type 10 to Type 251 Road Test Portfolio
Buick Performance Portfolio 1947-1962
Buick Muscle Portfolio 1963-1973
Buick Riviera Performance Portfolio 1963-1978
Cadillac Performance Portfolio 1948-1958
Cadillac Performance Portfolio 1959-1966
Cadillac Eldorado Performance Portfolio 1967-1978
Cadillac Allante Limited Edition Extra
Caterham Seven Road Test Portfolio 1974-1999
Caterham Seven Road Test Portfolio 2000-2010
Checker Automobiles Road Test Portfolio
Impala & SS Muscle Portfolio 1958-1972
Corvair Performance Portfolio 1959-1969
Chevy II & Nova SS Gold Portfolio 1962-1974
Chevelle & SS Gold Portfolio 1964-1972
Camaro Muscle Portfolio 1967-1973
Blazer & Jimmy Limited Edition Extra 1969-1982
Blazer & Jimmy Limited Edition Extra 1983-1994
Camaro Performance Portfolio 1993-2000
Chevrolet Caprice Road Test Portfolio 1965-1990
Chevrolet Corvair Performance Portfolio 1959-1965
Chevrolet Corvette Gold Portfolio 1953-1962
Chevrolet Corvette Sting Ray Gold Port. 1963-1967
Chevrolet Corvette Gold Portfolio 1968-1977
Chrysler Imperial Gold Portfolio 1951-1975
Chrysler 300 Gold Portfolio 1955-1970
Citroen Traction Avant Limited Edition Premier
Citroen 2CV Ultimate Portfolio 1948-1990
Citroen DS & ID 1955-1975
Citroen DS & ID Gold Portfolio 1955-1975
Citroen CX Road Test Portfolio
Citroen SM Limited Edition 1970-1975
Shelby Cobra Gold Portfolio 1962-1969
Crosley & Crosley Specials Limited Edition
Cunningham Automobiles 1951-1955
Daimler SP250 Dart & V-8 2.5 litre 250 RTP 1959-69
Datsun Roadsters Performance Portfolio 1960-71
Datsun 240Z & 260Z Road Test Portfolio
Datsun 280Z & 280ZX Road Test Portfolio
Delahaye Road Test Portfolio
Delage Road Test Portfolio
DeLorean Gold Portfolio 1977-1995
De Soto Limited Edition 1952-1960
De Tomaso Collection No. 1 1962-1981
Dodge Limited Edition 1949-1959
Dodge Dart Limited Edition Extra 1960-1976
Dodge Muscle Portfolio 1964-1971
Charger Muscle Portfolio 1966-1974
Edsel Limited Edition 1957-1960
Excalibur Collection No. 1 1952-1981
Elva Sports Racers Road Test Portfolio
ERA Gold Portfolio 1934-1994
Facel Vega Limited Edition Extra 1954-1964
Ferrari Road Cars 1946-1956 Road Test Portfolio
Ferrari Dino Limited Edition Extra 1965-1974
Ferrari 308 & Mondial Ultimate Portfolio 1975-85
Ferrari 328 348 Mondial Ultimate Portfolio 1986-94
Ferrari F355 & 360 Gold Portfolio 1995-2004

www.brooklandsbooks.com

Fiat 500 1936-1972 Road Test Portfolio
Fiat 600 & 850 Gold Portfolio 1955-1972
Fiat Dino Road Test Portfolio
Fiat 124 Spider Performance Portfolio 1966-1985
Fiat Abarth 1972-1987 Road Test Portfolio
Fiat X1/9 Gold Portfolio 1973-1989
Fiat Barchetta Road Test Portfolio
Ford Zephyr, Zodiac, Executive Mk. III & IV 1962-1971
High Performance Capris Gold Portfolio 1969-1987
Capri Muscle Portfolio 1974-1987
High Performance Fiestas 1979-1991
Ford Escort RS & Mexico Limited Edition 1970-1979
Ford Thunderbird Performance Portfolio 1955-1957
Ford Thunderbird Performance Portfolio 1958-1963
Ford Thunderbird Performance Portfolio 1964-1976
Ford Fairlane 1955-1970
Ford Fairlane Performance Portfolio 1955-1970
Ford Edsel 1957-1960 Road Test Limited Edition
Ford Ranchero Muscle Portfolio 1957-1979
Ford Falcon Performance Portfolio 1960-1970
Ford Galaxie and LTD Gold Portfolio 1960-1976
Ford GT40 & GT Ultimate Portfolio 1964-2006
Ford Bronco 4x4 Performance Portfolio 1966-1977
Ford Bronco 1978-1988
Ford Torino Performance Portfolio 1968-1974
Shelby Mustang Ultimate Portfolio 1965-1970
Mustang Muscle Portfolio 1967-1973
High Performance Mustang IIs 1974-1978
Mustang 5.0L Muscle Portfolio 1982-1993
Ginetta Cars Limited Edition Ultra 1958-2007
Goggomobil Limited Edition
Hispano-Suiza Road Test Portfolio
Holden 1948-1962
Honda S500, S600, S800 Limited Edition
Honda - Acura NSX Ultimate Portfolio 1989-2005
Honda CRX 1983-1987
Honda S2000 Performance Portfolio 1999-2008
Hudson Performance Portfolio 1946-1957
International Scout Gold Portfolio 1961-1980
Isetta Gold Portfolio 1953-1964
ISO & Bizzarrini Limited Edition Ultra 1962-1974
Jaguar XK120 XK140 XK150 Gold Portfolio
Jaguar Mk 7, 8, 10 & 420 Road Test Portfolio
Jaguar Mk 1 & Mk 2 1955-1969 Road Test Portfolio
Jaguar E-Type Ultimate Portfolio 1961-1975
Jaguar XJ6 Series I & II Gold Portfolio 1968-1979
Jaguar XJ6 Series III Perf. Portfolio 1979-1986
Jaguar XJ-S V12 Ultimate Portfolio 1988-1996
Jaguar XK8 & XKR Performance Portfolio 1996-2005
Jeep CJ-5 Limited Edition 1960-1975
Jeep CJ-5 & CJ-7 4x4 Perf. Portfolio 1976-1986
Jeep Wagoneer Performance Portfolio 1963-1991
Jeep J-Series Pickups 1970-1982
Jeepster & Commando Limited Edition 1967-1973
Jeep Cherokee & Comanche Pickups P. P. 1984-1991
Jeep Wrangler 4x4 Performance Portfolio 1987-1999
Jeep Cherokee & Grand Cherokee 4x4 P. P. 1992-1998
Jensen Road Test Portfolio 1934-1965
Jensen Interceptor Ultimate Portfolio 1966-1992
Jensen -Healey Road Test Portfolio 1972-1976
Jowett Road Test Portfolio
Kaiser -Frazer Limited Edition 1946-1955
Lagonda Gold Portfolio 1919-1964
Lancia Aurelia & Flaminia Gold Portfolio 1950-1970
Lancia Fulvia Gold Portfolio 1963-1976
Lancia Montecarlo & Scorpion 1975-1982 RTP
Lancia Stratos Limited Edition Extra
Lancia Delta & integrale Ultimate Portfolio
Land Rover Series I, II & IIA Gold Portfolio 1948-1971
Land Rover 90 110 Defender Gold Portfolio 1983-1994
Lamborghini Performance Portfolio 1964-1976
Lamborghini Performance Portfolio 1977-1989
Lamborghini Gold Portfolio 1990-2004
Lincoln Gold Portfolio 1949-1960
Lincoln Continental Performance Portfolio 1961-1969
Lincoln Continental 1969-1976
Lola Sports Racing Cars Limited Ed. Premier 1958-1985
Lotus Sports Racers Portfolio 1951-1965
Lotus Seven Gold Portfolio 1957-1973
Lotus Elite Limited Edition 1957-1964
Lotus Elan Ultimate Portfolio 1962-1974
Lotus Cortina Road Test Portfolio
Lotus Elite & Eclat Road Test Portfolio 1984-1972
Lotus Excel Road Test Portfolio
Lotus Europa Gold Portfolio 1966-1975
Lotus Elise Limited Edition 1996-1999
Lotus Elise & Exige Gold Portfolio 1996-2005
Marcos Coupés & Spyders Gold Portfolio 1960-1997
Maserati Cars Performance Portfolio 1957-1970
Maserati Cars Performance Portfolio 1971-1982
Maserati Cars Performance Portfolio 1982-1998
Maserati Cars Ultimate Portfolio 1999-2007
Matra Road Test Portfolio
Mazda Miata MX-5 Performance Portfolio 1989-1997
Mazda Miata MX-5 Performance Portfolio 1998-2005
Mazda RX-7 Gold Portfolio 1979-1992
McLaren F1 • GTR • LM Sportscar Perf. Portfolio
Mercedes-Benz 1925-1939 - A Portfolio
Mercedes 190SL 300SL 300SLR - A Portfolio
Mercedes S & 600 Limited Edition Extra 1965-1972
Mercedes S Class Limited Edition Extra 1980-1991
Mercedes 230SL 250SL 280SL Ultimate Port. 1963-1971
Mercedes-Benz SLs & SLCs Ultimate Port. 1971-1989
Mercedes SLs Performance Portfolio 1989-1994
Mercedes E-Class W124 Road Test Portfolio 1986-1995
Mercedes G-Wagen Gold Portfolio 1981-2005
Mercedes AMG Gold Portfolio 1983-1999
Mercedes AMG Ultimate Portfolio 2000-2006
Mercury Gold Portfolio 1947-1966
Mercury Comet & Cyclone Lim. Edit. Extra 1960-1975
Mercury Cougar Muscle Portfolio 1967-1973
Messerschmitt Gold Portfolio 1954-1964

MG Gold Portfolio 1929-1939
MG TD & TF Gold Portfolio 1949-1955
MG Y Type & Magnette Road Test Portfolio
MGA & Twin Cam Gold Portfolio 1955-1962
MG Midget Road Test Portfolio 1961-1979
MGB Roadsters 1962-1980
MGB MGC & V8 Gold Portfolio 1962-1980
MGF & TF Performance Portfolio 1995-2005
Mini Gold Portfolio 1959-1969
Mini Gold Portfolio 1969-1980
Mini Gold Portfolio 1981-1997
High Performance Minis Gold Portfolio 1960-1973
Mini Cooper Gold Portfolio 1961-1971
Mini Moke Ultimate Portfolio 1964-1994
Starion & Conquest Performance Portfolio 1982-1990
Mitsubishi 3000GT & Dodge Stealth P.P. 1990-1999
Morgan Three-Wheelers Ultimate Portfolio 1909-52
Morgan Four-Wheelers Ultimate Portfolio 1936-67
Morgan Ultimate Portfolio 1968-1990
Morgan Ultimate Portfolio 1991-2009
Morris Minor MM & Series II Road Test Portfolio
Morris Minor 1000 Road Test Portfolio
Nash Limited Edition Extra 1949-1957
Nash-Austin Metropolitan Gold Portfolio 1954-1962
Nissan 350Z Road Test Portfolio
Nissan Skyline GT-R Ultimate Portfolio 1969-2010
Nissan Skyline GT-R Limited Edition Extra 1989-2002
Noble Sports Cars Road Test Portfolio
NSU Ro80 Limited Edition
Oldsmobile 1948-1963 Limited Edition Premier
Oldsmobile Cutlass & 4-4-2 Muscle Port. 1964-1974
Oldsmobile Muscle Cars 1964-1971
Oldsmobile Toronado 1966-1978
Opel GT Ultimate Portfolio 1968-1973
Opel Manta Limited Edition 1970-1975
Packard Automobiles 1920-1958 - A Portfolio
Pantera and Mangusta 1969-1974
Pantera Ultimate Portfolio 1970-1995
Panther Gold Portfolio 1972-1990
Plymouth Limited Edition 1950-1960
Plymouth Fury Limited Edition Extra 1956-1976
Barracuda Muscle Portfolio 1964-1974
Plymouth Muscle Portfolio 1964-1971
Pontiac 1946-1963 Limited Edition Premier
High Performance Firebirds 1982-1988
Firebird & Trans Am Performance Portfolio 1993-00
Pontiac Fiero Performance Portfolio 1984-1988
Porsche Sports Racing Cars UP 1952-1968
Porsche 917 • 935 • 956 • 962 Gold Portfolio
Porsche 365 Ultimate Portfolio 1952-1965
Porsche 911 1965-1969
Porsche 911 SC & Turbo Gold Portfolio 1978-1983
Porsche 911 Ultimate Portfolio 1990-1997
Porsche 911 Ultimate Portfolio 1998-2004
Porsche 912 Limited Edition Extra 1965-1976
Porsche 914 Ultimate Portfolio
Porsche 924 Gold Portfolio 1975-1988
Porsche 928 Gold Portfolio 1977-1995
Porsche 928 Takes On The Competition
Porsche 944 Ultimate Portfolio
Porsche 968 Limited Edition Extra
Porsche Boxster Ultimate Portfolio 1996-2004
Railton & Brough Superior Gold Portfolio 1933-1950
Range Rover Gold Portfolio 1970-1985
Range Rover Gold Portfolio 1985-1995
Range Rover Performance Portfolio 1995-2001
Riley RM Series Pathfinder 2.6 A Brooklands Port.
Rolls-Royce Silver Cloud & Bentley S Ultimate Port.
Rolls-Royce Silver Shadow Ultimate Port. 1965-80
Rover P4 1949-1959
Rover 3 & 3.5 P5 Gold Portfolio 1958-1973
Rover P6 1963-1977
Rover 2000 & 2200 1963-1977
Saab Sonett Collection No. 1 1966-1974
Studebaker Ultimate Portfolio 1946-1966
Studebaker Hawks & Larks Limited Edit. Premier 1956-66
Avanti Limited Edition Extra 1962-1991
Subaru Impreza WRX Performance Port. 2001-2005
Sunbeam Alpine and Tiger Gold Portfolio 1959-1967
Sunbeam Alpine Limited Edition Extra 1959-1968
Sunbeam Tiger Limited Edition Extra 1964-1967
Suzuki SJ Gold Portfolio 1971-1997
Tatra Cars Road Test Portfolio
Toyota Land Cruiser Gold Portfolio 1956-1987
Toyota Land Cruiser 1988-1997
Toyota Supra Performance Portfolio 1982-1998
Toyota MR2 Road Test Portfolio 1984-1989
Toyota MR2 Road Test Portfolio 1990-1999
Toyota MR2 Road Test Portfolio 2000-2007
Triumph TR2 - TR3 1952-1960
Triumph TR2 & TR3 Gold Portfolio 1952-1961
Triumph TR4, TR5, TR250 1961-1968
Triumph TR6 Road Test Portfolio
Triumph TR7 & TR8 1975-1982
Triumph TR7 TR8 Gold Portfolio 1975-1982
Triumph Herald 1959-1971
Triumph Vitesse 1962-1971
Triumph 2000, 2.5, 2500 1963-1977
Triumph GT6 Gold Portfolio 1966-1974
Triumph Spitfire Road Test Portfolio
Triumph Stag Road Test Portfolio
TVR Limited Edition Ultra 1958-1985
TVR Performance Portfolio 1986-1994
TVR Performance Portfolio 2000-2005
VW Beetle Gold Portfolio 1935-1967
VW Beetle Gold Portfolio 1968-1991
VW Bus Camper Van Perf. Portfolio 1954-1967
VW Bus Camper Van Perf. Portfolio 1968-1979
VW Bus Camper Van Perf. Portfolio 1979-1991
VW Karmann Ghia Gold Portfolio 1955-1974
VW Golf GTI Limited Edition Extra 1976-1991
VW Golf Cabriolet 1979-2002 Road Test Portfolio
VW Corrado Limited Edition Premier 1989-1995

Volvo PV444 & PV544 Perf. Portfolio 1945-1965
Volvo 120 Amazon Ultimate Portfolio
Volvo 1800 Ultimate Portfolio 1960-1973
Volvo 140 & 160 Series Gold Portfolio 1966-1975
Westfield Performance Portfolio 1982-2004

MILITARY VEHICLES

Complete WW2 Military Jeep Manual
Dodge WW2 Military Portfolio 1940-1945
German Military Equipment WW2
Hail To The Jeep
Combat Land Rover Portfolio No. 1
Land Rover Military Portfolio
Off Road Jeeps Civilian & Military 1944-1971
US Military Vehicles 1941-1945
Standard Military Motor Vehicles-TM9-2800 (WW2)
VW Kubelwagen Military Portfolio 1940-1990
WW2 Allied Vehicles Military Portfolio 1939-1945
WW2 Jeep Military Portfolio 1941-1945

RACING & THE LAND SPEED RECORD

The Land Speed Record 1898-1919
The Land Speed Record 1920-1929
The Land Speed Record 1930-1939
The Land Speed Record 1940-1962
The Land Speed Record 1963-1999
Can-Am Racing 1966-1969
Can-Am Racing 1970-1974
Can-Am Racing Cars 1966-1974
The Carrera Panamericana Mexico - 1950-1954
Le Mans - The Bentley & Alfa Years - 1923-1939
Le Mans - The Jaguar Years - 1949-1957
Le Mans - The Ferrari Years - 1958-1965
Le Mans - The Ford & Matra Years - 1966-1974
Le Mans - The Porsche Years - 1975-1982
Le Mans - The Porsche & Jaguar Years - 1983-91
Le Mans - The Porsche & Peugeot Years - 1992-99
Mille Miglia - The Alfa & Ferrari Years - 1927-1951
Mille Miglia - The Ferrari & Mercedes Years - 1952-57
Targa Florio - The Ferrari & Lancia Years - 1948-54
Targa Florio - The Porsche & Ferrari Years - 1955-64
Targa Florio - The Porsche Years - 1965-1973
Brabham Ralt Honda The Ron Tauranac

RESTORATION & GUIDE SERIES

BMW 2002 - A Comprehensive Guide
BMW 02 Restoration Guide
BMW E30 - 3 Series Restoration Bible
BMW E36 - 3 Series Restoration Tips & Techniques
BMW - 5 & 6 Series Restoration Tips & Techniques
Classic Camaro Restoration
Land Rover Restoration Portfolio
PC on Land Rover Series I Restoration
Lotus Elan Restoration Guide
Lotus Twin Cam Engines
MG T Series Restoration Guide
MGA Restoration Guide
PC on MGB Restoration
Mustang Restoration Tips & Techniques
Practical Gas Flow
Range Rover - The First Fifty
Restoring Sprites & Midgets an Enthusiast's Guide
Solex Carburettors Tuning Tips and Techniques
SU Carburetters Tuning Tips & Techniques
The Great Classic Muscle Cars Compared
Weber Carburettors Tuning Tips and Techniques

ROAD & TRACK SERIES

Road & Track on Aston Martin 1962-1990
Road & Track on Auburn Cord & Duesenberg 1952-1984
Road & Track on Austin Healey 1953-1970
R & T BMW Z3, M Coupe & M Roadster Port. 1996-02
R & T Camaro & Firebird Portfolio 1993-2002
R & T on Cobra, Shelby & Ford GT40 1962-1992
Road & Track Dodge Viper Portfolio 1992-2002
Road & Track Ferrari V-12 Portfolio 1992-2002
Road & Track Ferrari F355 360 F430 Portfolio 1995-06
R & T Jaguar XJ-S - XK8 - XKR Portfolio 1975-2003
Road & Track MX-5 Miata Portfolio 1989-2002
R & T Mercedes SL - SLK - CLK Portfolio 1990-2003
Road & Track on MG Sports Cars 1949-1961
Road & Track on MG Sports Cars 1962-1980
Road & Track Mustang Portfolio 1994-2002
Road & Track Nissan 300ZX & 350Z Portfolio 1984-03
Road & Track Porsche 928 Portfolio 1977-1994
Road & Track Porsche 911 Portfolio 1990-1997
Road & Track on Triumph Sports Cars 1953-1967
Road & Track on Triumph Sports Cars 1967-1974
Road & Track on Triumph Sports Cars 1974-1982
Road & Track - Henry Manney at Large and Abroad
Road & Track - Peter Egan Side Glances 1983-1992
Road & Track - Peter Egan Side Glances 1992-1997
Road & Track - Peter Egan Side Glances 1998-2002
Road & Track - Peter Egan Side Glances 2002-2006
Road & Track Best of PS

CAR AND DRIVER SERIES

Car and Driver on Cobra, Shelby and Ford GT40 1963-84

ENGINE SERIES

Chevy 265 & 283
Chevy 302 & 327
Chevy 348 & 409
Chevy 350 & 400
Chevy 396 & 427
Chevy 454 thru 512
Chrysler Hemi
Chrysler 273, 318, 340 & 360
Chrysler 361, 383, 400, 413, 426 & 440
Ford 289, 302, Boss 302 & 351W
Ford 351C & Boss 351
Ford Small Block
Ford Big Block

MOTORCYCLES

To see our range of over 70 titles visit www.brooklands-books.com

CONTENTS

ACKNOWLEDGEMENTS

When buyers of our first book about Alfa Romeo coupés of the 1970s and 1980s asked for more material about the GTV6, we set about searching through our archives to see what we could find. The results are incorporated in this new book, which we have devoted exclusively to these high-performance coupés. None of the material we have included here duplicates that in our Alfa Romeo Alfetta Gold Portfolio, although GTV6 fans may wish to look there for further material about their favourite cars.

Our archives, and our books, can only be as good as the material available to us. So we are delighted to express out thanks here to the holders of the copyright to the articles as follows. It is through their kind co-operation that we are able to reproduce material originally published in Asian Auto, Autocar, Autosport, Car, Car and Driver, Car South Africa, Fast Lane, Modern Motor, Motor, Motor Manual, Motor Sport, Popular Classics, Road & Track, Sporting Cars, Sports Car World, Thoroughbred & Classic Cars, What Car? and Your Classic.

R. M. Clarke

For the styling of the coupé companion to their Alfetta saloon, Alfa Romeo turned to Giugiaro. There can be no doubt that the wedge-shaped hatchback design he put forward was one of the classic shapes of the 1970s, and by the end of the decade it still appeared fresh and arresting.

It was in 1980 that Alfa Romeo put their new 2.5 litre 60-degree V5 engine into the GT coupé, renaming it a GTV6 and offering it alongside the existing four-cylinder 2-litre GTV model. The GTV6 was undoubtedly the crowning achievement of the range, combining proven qualities of handling and balance with the urge of a 160bhp engine. The combination produced maximum speeds of just under 130mph, although to achieve 0-60 mph in the 8.2 seconds sometimes claimed for the cars took practice; the gear change linkage to the rear-mounted transaxle did not encourage slick and swift changes, either up or down the 'box.

Nevertheless, the GTV6 developed into a successful competition car in the early 1980s, scoring well in touring car races and taking Andy Rouse to the 1984 European Championship. Unlike many of the earlier Alfa coupés, the GTV6s were fairly well protected against corrosion, with the result that there is no shortage of examples still around for enthusiasts to enjoy. And, as this book makes clear, there is a lot to enjoy about a GTV6!

James Taylor

WHILE OTHER manufacturers have played safe with the committee-conceived box, Alfa Romeo have consistently produced cars that offer something special in design and style.

The GTV's advanced chassis specification would seem reason enough for giving the car more power and installing the 2.5-litre V6 (introduced last year in the Alfa 6 saloon). However Alfa cite the *Raison d'etre* as being an upturn in demand for their sports cars, particularly in the US; a demand that can hardly be hindered by their recent ultra competitive performance in formula 1. At this point it is worth mentioning that the Alfa/Nissan,

An angular bonnet hump distinguishes the GTV6 2.5. Alloy wheels carry P6 tyres. Moulded spoiler extends upwards to cover front bumper. Rear bumper (below) is also capped by plastic moulding Far right: The GTV 2.0 is recognised by its smooth bonnet line and steel wheels

Bravo Alfa!

Alfa's 2.5 litre V6 appears in facelifted GTV range

By John Miles

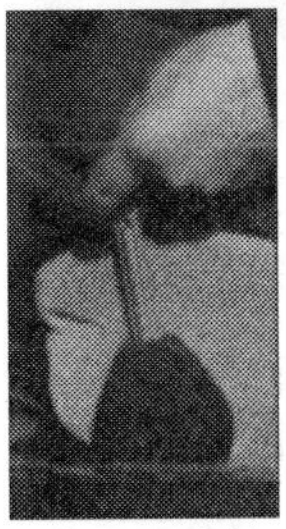

An extendable lever provides adequate leverage on the seat height adjustment

a car designed to take up Alfa's overcapacity in production of mechanical parts by fitting them in a Nissan body (the bare panels are assembled by Alfa), should be in production within three years.

The updated GTV 2.0 remains mechanically unchanged. Alfa's extraordinarily long-lived double overhead camshaft engine gives 130 bhp at 5,400 rpm and 132 lb ft torque at 4,000 rpm. As on the Alfetta saloon and Giulietta the drive goes back to a five-speed all-indirect transaxle sitting between the rear wheels; the clutch is also rear mounted. Close to 50/50 weight distribution results. Front suspension is by double wishbones (torsion bar sprung). At the rear there is the usual Alfa de Dion axle (coil sprung) located longitudinally by long fixed radius arms converging to a fixing point on a cross member. Lateral location is by Watt linkage. Telescopic dampers are used all round, and there are anti-roll bars front and rear.

Main interest centres on the GTV6 2.5. With an 88mm bore and 68.3mm stroke Alfa's all-alloy, wet-linered, 60 deg V6 (fully described in *Autocar* 28 April 1979) is significantly oversquare. Without resorting to double overhead camshafts, Alfa achieve near hemispherical combustion chambers with a single camshaft per bank, by working the inlet valves directly through inverted buckets, and the exhausts via a transversely placed pushrod which operates a rocker.

As used in the GTV Alfa's 2.5-litre V6 employs Bosch L Jetronic fuel injection. Aluminium is employed for heads, block and sump

So as to get the engine to conform to current US and Swedish toxic emission standards, the six individual downdraught Solex carburettors found on the Alfa 6 saloon have been eschewed on the GTV6 in favour of Bosch L Jetronic electronic fuel injection. On a moderate 9.0 to 1 compression ratio peak power output is unchanged at 160 bhp, but is developed at 6,000 rpm instead of at 5,800 rpm as in the saloon. Maximum torque on the other hand occurs at 4,000 rpm (the same) but at 157 lb ft is some 6lb ft less than on the carburettor'd motor. Alfa quote kerb weight at 23.9 cwt (1.9 cwt more than the GTV 2.0) but some 5 cwt less than the Alfa Six saloon, so the small sacrifice in torque is of little consequence.

Transmission modifications are extensive. The standard propshaft carries larger couplings. In place of the GTV 2.0's single 215mm dia clutch, the GTV 2.5 is fitted with a twin plate of the same diameter (a single plate of the required size would not have fitted within the bellhousing) The GTV 2.5's 23 per cent greater power and 18 per cent torque improvement have required the existing transaxle to be fitted with modified gears and larger bearings. The thrust bearing and clutch master cylinder are also new.

When questioned as to why the GTV 2.5 runs a lower ratio first gear than the GTV 2.0, (even when the latter's is on the low side), an Alfa spokesman rather side-stepped the issue by saying the lower first gave "easier starting characteristics from rest". Fifth gear is raised to give overall gearing of 21.9 mph per 1,000 rpm —virtually ideal to achieve the claimed (and believable) 128 mph maximum speed at the 6,000 rpm power peak.

Other parts peculiar to the big-engined GTV are its ventilated and slightly larger diameter front discs (GTV 2.0 rear discs remain) and 16.6 gallon fuel tank. The car is shod with 195/60 Pirelli P6 tyres mounted on 15 x 6in. Campagnolo alloy wheels. Suspension settings have been revised (upwards) to take advantage of the much improved levels of grip.

From the outside the new GTVs take on an altogether less fussy look. The 2.6 is easily distinguished by its angular bonnet hump made necessary to give clearance over the fuel injection plenum. Brightwork is now virtually absent. A deep full width moulded front spoiler replaces the previous bib type, and this moulding extends upwards to cap the front bumper. It contains inset indicator lamps. The rear bumper is similarly capped and there are now side sill protection strips moulded in the same dark grey material. The theme continues with black satin finish (baked epoxy) on the window frames and black rubber strips neatly covering the guttering. The car looks cleaner, but Alfa say no measureable improvement in aerodynamic efficiency results. A drag co-efficient of 0.39 is quoted.

Inside, the most welcome and significant improvement is to the facia. The main instruments are now grouped together conventionally (as opposed to the previously fragmented layout) in a single hooded binnacle. A centre console panel houses the ancilliary gauges, warning lights, and switchgear. The driving seats have a proper parallelogram type height adjustment (rather than the more commonly found tipping type). And there is a particularly neat extendable lever which allows height adjustment without the driver having to ease his weight off the seat. In other respects the fittings are unchanged. As before the seats and door panels are trimmed in ribbed velour.

GTV 2.0s and 2.5s will come finished in black, red, or grey. The right hand drive version is scheduled to appear here in May 1981, and Alfa indicate the prices will be approximately £9,250, and £10,200 respectively.

Driving Impressions

Would that Alfa could have brought themselves to make the steering column adjustable for reach (there is the usual rake adjustment). We have to compromise and sit with legs a little bent so as to be able to reach "over the top" of the wheel. Likewise the gearlever is a little too distant when changing from second to third or from fourth to the dog leg fifth; a gear lever movement that should become far more natural on the right hand drive cars. Any initial feeling of the seats being too firm soon disappears. They are comfortable and will provide us (my driving partner is ex-*Autocar* staff man "Steady" Barker) with excellent support.

Even while trickling out towards the country via Rome's crowded suburbs the GTV 2.5

gives the driver the feeling that he is conducting a piece of well-oiled machinery. The roads begin to thin of traffic, and become twistier. The GTV 2.5 begins to *live*. The engine pulls strongly from low rpm with a complete lack of temperament. At just over 4,000 rpm induction chatter is joined by healthy exhaust rasp. The revs build willingly and silkily to the 6,000 rpm red line. Such music is muted, yet under hard acceleration it is always there to call the driver on.

It's a real Alfa, with arrowlike straight line stability (ours had) and a wonderful feeling of swervability. Only truly crude treatment (braking in mid corner) or too much throttle out of a slow and slimy corner would unstick the rear. Such rear end grip makes it a delightful car to drive hard into a corner under braking, or when racking it from left to right through an S-bend.

Ultimately, rear end adhesion *is* better than at the front yet not to the imbalanced degree it is on the Guilietta – a dramatic understeerer. Nor does the car have the Porsche 924's rather rubbery initial steering response. By comparison, the steering is razor sharp, and has excellent feel, (it lightens perceptibly as we traverse a damp patch in mid corner) and is reasonably geared (3.5 turns). Steering loads build up nicely in relation to cornering speeds.

In fast open bends the car inspires terrific confidence. At *going* speeds, i.e. those where one can see to stop, we always find reserves of front-end adhesion in the slower corners to haul the front into a tightening radius. Lifting off in mid corner neutralises the tendency for the car to run wide if the entry has been a shade over-enthusiastic.

This taut handling feel – togetherness – has been achieved with a minimal cost in ride (a strong point on more prosaic Alfas). Surfaces equating to the very worst British B road provoke a knobbly ride but that is to be expected in a P6-shod sports car, and it is noticeable that when the suspension is "loaded" in mid corner the car runs more smoothly over the rough stuff.

In most respects ride quality can hardly be faulted (although one uneven section of motorway did excite a sympathetic bumpiness in the car.) Ride is taut yet supple over most surfaces and the car takes long undulations beautifully. The brakes are averagely light to work, progressive and powerful without being spongey.

Alfa appear to have done particularly well in subduing road-induced noises – all too difficult with low profile tyres. Rumble and thump are there but never to an obtrusive degree, and we were particularly impressed by the virtual absence of pattern noise (the roar of varying intensity which low profile tyres are prone to generate particularly on coarser surfaces).

Judging from the ease with which we saw an indicated 210 kph (130 mph) there is no reason to doubt Alfa's claimed top speed. You can sit back and cruise it between 110-120 mph. Most of the exhaust noise is left behind. There is subdued induction chatter, and a moderate degree of wind noise. Normal conversation is possible.

A near perfect machine then? Not quite. A rather ponderous gearchange between the lower gears inhibits keen driving on twisty roads and please Alfa Romeo can we have more sensible first and second gear ratios? They are good for 30 and 50 mph respectively (40 and 60 mph would befit the car). There is a huge 3,600 rpm rev drop between the two, and 1,900 rpm between second and the 76mph third gear. Quick selection of first while at rest is accompanied by a crunch that suggests weak synchromesh on that gear – a characteristic found on other current transaxled Alfas.

Our particular test car also seemed to suffer from lacklustre ram fed fresh airflow. Adequate ventilation was only possible with fan assistance (noisy) when the facia top vents could be brought into action.

Into sun, the GTV's steeply sloping screen is prone to suffer facia top reflections, and one notices the steeply raked pillars inhibiting one's sightline in town and when looking "across" a corner.

Niggles apart (please can those gear ratios be re-arranged?) it still seemed that installing the 2.5-litre V6 in the GTV has done nothing but turn a likeable car into a delightful one.

Left: Main instruments and clock are now grouped in one binnacle. Fuel contents, oil pressure and water temperature gauges plus ancilliary warning lights and switchgear sit in centre console panel. A new steering wheel incorporates centre horn press

SPECIFICATION

	GTV6 2.5	GTV 2.0
ENGINE	Front, rear drive	Front, rear drive
Head/block	Alloy/alloy	Alloy/alloy
Cylinders	6 in 60 deg V	4
Main bearings	4	5
Cooling	Water	Water
Fan	Twin electric	Electric
Bore, mm (in.)	88 (3.46)	84 (3.31)
Stroke, mm (in.)	68.3 (2.69)	88.5 (3.48)
Capacity, cc (in³)	2,492 (152.3)	1,962 (119.7)
Valve gear	Ohc	Dohc
Camshaft drive	Cogged belt	Chain
Compression ratio	9.0 to 1	9.0 to 1
Ignition	Electronic	Contact breaker
Fuel injection	Bosch L Jetronic	Twin Dellorto twin-choke
Max power	160 bhp at 6,000 rpm	130 bhp at 5,400 rpm
Max torque	157 lb ft at 4,000 rpm	132 lb ft at 4,000 rpm

TRANSMISSION

	GTV6 2.5 Ratio	mph/1000rpm	GTV 2.0 Ratio	mph/1000rpm
Type	5 speed all indirect transaxle			
Clutch	Twin disc diaphragm			
Gear	Ratio	mph/1000rpm	Ratio	mph/1000rpm
Top	0.780	21.9	0.830	20.8
4th	1.026	16.7	1.026	16.9
3rd	1.345	12.7	1.345	12.8
2nd	1.956	8.7	1.956	8.8
1st	3.500	4.9	3.300	5.2
Final drive gear	Hypoid		Hypoid	
Ratio	4.1		4.1	

	GTV6 2.5	GTV 2.0
SUSPENSION		
Front – location	Double wishbones	
springs	Longitudinal torsion bars	
dampers	Telescopic	
anti-roll bar	Yes	
Rear – location	De Dion tube, Watt linkage	
springs	Coil	
dampers	Telescopic	
anti-roll bar	Yes	
STEERING		
Type	Rack and pinion	
Power assistance	No	
Turns lock to lock	3.5	
BRAKES		
Circuits	Dual	
Front	10.5 in. dia. ventilated disc	10.3 in. dia. disc
Rear	9.8 in. dia. disc	9.8 in. dia. disc
Servo	Vacuum	
Handbrake	Centre lever working on rear brakes	
WHEELS		
Type	Alloy	Pressed steel
Rim width	6 in.	5½ in.
Tyres – make	Pirelli	Various
type	P6	Radial ply
size	195/60-15	185/70 HR14
EQUIPMENT		
Battery	12V 77Ah	12V 50Ah
Alternator	65A	45A
Headlamps	220/110 Watt (total)	
Reversing lamp	Standard	
Hazard warning	Standard	
Screen wipers	2-speed + intermittent	
Screen washer	Electric	
Interior heater	Air blending	
Air conditioning	Extra	
Interior trim	Cloth seats, pvc headlining	
Floor covering	Carpet	
Windscreen	Laminated	
Underbody protection	Zinc and cadmium plating, plastic coating, wax in box sections	
DIMENSIONS		
Wheelbase	94.5 in. (2,400 mm)	
Track front	54 in. (1,373 mm)	
rear	53 in. (1,352 mm)	
Overall length	168 in. (4,260 mm)	
width	65.5 in. (1,664 mm)	
height	52.5 in. (1,330 mm)	
Kerb weight DIN standard	2,664 lb (1,210 kg)	
SERVICE DATA		
Fuel tank	16.6 Imp. galls (75.5 litres)	11.9 galls (54 litres)
Engine sump	20 pints (12 litres)	
Gearbox oil inc final drive	4.9 pints (2.8 litres)	

PERFORMANCE (Mfrs figures)	**GTV6 2.5**	**GTV 2.0**
Maximum speed (approx.)	128 mph	118 mph
Standing km	29.5 sec	30.3 sec
ECE/Government Official Fuel Consumption figures		
Urban (manual/automatic)	19.1 mpg (14.8 litres/100km)	20.4 mpg 13.9 litres/100km
Steady 56 mph (90 kph)	36.7 mpg (7.7 litres/100km)	40.9 mpg 6.9 litres/100km
Steady 75 mph (120 kph)	28.5 mpg (9.9 litres/100km)	29.4 mpg 9.6 litres/100km

Alfa Romeo GTV 6 2.5 iniezione

AFTER a short, but very testing, drive in the latest car of the Alfa Romeo GTV series, I would recommend it to anyone with about £10,000 to spend wanting a sporty, occasional four-seater. It is a delight to drive and is every bit an Alfa, right down to the crisp and very Italian exhaust note emanating from the 2.5-litre V6 engine.

The external appearance follows that of the existing GTV range very closely, but with detail refinements and improvements such as slow impact absorbing polypropylene covered bumpers front and rear, a bib spoiler at the front, altered lamp clusters at the rear, a black moulding along the lower edge of the body linking the wheel arches, to protect the sills, black opaque window frames and a power bulge in the bonnet. This latter is necessary to give clearance for the new engine, and is about the only unsatisfactory part of the car — to me, at least, it foreshortens the bonnet and upsets the previously sleek appearance. Others may prefer to describe it as "blunt and purposeful", or "aggressively attractive" — I just think it spoils the shape of a very elegant car. The wheels are of light alloy, made to an Alfa Romeo design by Campagnolo, and are fitted with Pirelli P6 tyres.

Inside, the trim is improved and purchasers may opt for either velour or leather facings. The seating arrangement remains much the same, with the driver's seat being adjustable for height by means of Alfa's ingenious lever and cam arrangement, the access to the rear seats being assisted by the way the whole of the front seat slides forward when the back is tipped down. The rear seats are comfortable, but are restricted in size, and would not suit large people, or even medium sized people if the front seat passengers were being selfish over legroom: for the man with two children in their early teens, it would be ideal. The luggage space, however, is not so good. The larger capacity engine has dictated a larger fuel tank; this is situated behind the rear seats and encroaches significantly into the luggage area. The remaining space will accommodate far more than a couple of toothbrushes, and would be adequate for most holidaying couples, but not for the man with his wife and two teenagers.

The steering wheel has three wide spokes emanating from a large circular centre pad which incorporates the horn push and has a polished wooden rim. Personally, I am not keen on wooden rimmed wheels as they tend to become slippery in hot conditions, but this one is very elegant, and a pleasure to handle. The column is adjustable for height, but not for reach, which is a pity as I would have liked to have had the wheel about half an inch closer to the seat for the perfect position.

Undoubtedly the most noticeable alteration inside the car is the completely revised instrument panel. Gone are the separate consoles for speedometer and tachometer. In their place is a wide panel, well shielded, with a crushable moulding, containing three large and very clear instruments (speedometer, tachometer and clock) directly in front of the driver, with the secondary instruments, switches and warning lights grouped together over the centre console but still within the crushable moulding. The heating and ventilating controls come below this, directly above the space for a radio. Wipers and lights are stalk operated and there is a dash mounted cold starting control. The centre console reaches back to between the front seats, the area between the radio and the short wooden-knobbed gear lever being carpeted for use as a capacious oddments tray — this being particularly useful as the glove box (not lockable) and the map tray above the passenger's feet are both of rather limited size. Behind the gear lever, and alongside the hand brake lever, are a cigarette lighter and the control for the electrically operated driver's door mirror — something of an unnecessary luxury, I feel, as it is very easy to adjust the driver's mirror manually from the driving seat. Mercedes have the right idea with their new S Class saloons where the passenger's door mirror is electrically operated, while the driver's is manually adjustable from within the car. The pedals are well spaced, and are light to control and a rest is provided for the left foot.

The engine is the same basic unit as that used in the Alfa 6 saloon, but in fuel injected form. Apparently, Alfa Romeo considered using fuel injection on the saloon, but found that the performance in the lower engine speed range was inferior to the carburetter version and affected the flexibility of the larger car. With the lighter and sleeker shape of the GTV, and its more sporting character, low engine speed flexibility is no longer a problem and the benefits of Bosch L-Jetronic fuel injection can be enjoyed — for a start, this well tried system must be simpler to set up and maintain than the six individual carburetters fitted to the Alfa 6, and the fuel consumption and exhaust emission are both improved. The compression ratio is 9:1, the same as the Alfa 6, and maximum power is quoted as 160 b.h.p. as 6,000 r.p.m. — the Alfa 6 is quoted at 160 b.h.p. at 5,800 r.p.m. It is a beautifully smooth engine, which revs extremely easily making the automatic rev. limiter in the distributor of the breakerless ignition system a useful fitting — it operates at just over 6,100 r.p.m. The nitrided crankshaft runs in four main bearings, and, as would be expected from a very oversquare (88 × 68.3 mm.) and compact 60° V6 engine, is very short and stiff. With the low compression ratio, this engine should have great potential for the tuning enthusiasts. Crankcase/block and cylinder head are of light alloy construction and the very tall but broad heads contain a single belt driven o.h. camshaft which operates the inlet valves directly and the sodium-filled exhaust valves by way of short pushrods and rockers.

To cope with the extra power of the V6 engine, the transmission has been uprated. The propshaft, which runs at engine speed the whole time as the clutch is mounted at the back of the car, has been beefed up, and a two disc clutch has been fitted because the required increase in diameter of the single plate clutch fitted to the current GTV would have meant altering the rear floor panels to clear a larger housing, resulting in reduced headroom for the rear seat passengers. The five-speed gearbox has a lower bottom gear and a higher top gear than the current 2000 GTV, but the other ratios are identical, as is the final drive ratio. I suspect that bottom gear is lower to comply with legislation making it essential for the car to be able to re-start on the steepest Alpine road when fully laden and towing a trailer of the maximum specified weight — it could usefully be half a ratio higher. The rear suspension is of the de Dion type, with coil springs, the front is by torsion bars. Telescopic hydraulic shock absorbers are fitted all round and there are anti-roll bars front and rear — just as on the current GTV. Steering is rack and pinion, 3½ turns lock to lock, and the brakes are disc all round, servo-assisted.

Our test route took us up into the hills to the north-west of Rome, before looping round onto the Autostrada for the return to the capital. The roads were very well chosen for their variety and character, and we were able to assess the ride on poor roads, the handling on mountain lanes, the roadholding on wide open curves, the acceleration when passing other motorists, fast cruising speeds

on the Autostrada — and in all respects, the GTV 6 came out well. The steering was light and precise and totally in keeping with the Alfa Romeo tradition of near perfection in this respect. There had been talk of fitting the car with power steering to cope with the 195 × 15 60% profile P6 Pirelli tyres at low speeds, but this would have completely wrecked the beautiful balance achieved with the unassisted set up — and the steering is not heavy at low speeds anyway; only those who use power assistance to jack the car away from kerbs will miss it. The brakes were good, but with rather a long travel to the pedal and without the superb assurance of those of the Alfa 6.

The clutch control is very light and progressive. Bottom gear, as has been mentioned, seems somewhat low under normal conditions taking the car up to 30 m.p.h. at maximum r.p.m. The change from first to second is light, but not fast, and second goes to just over 50 m.p.h. Maximum revs in third and fourth represent slightly over 75 and 100 m.p.h. respectively, while fifth, on a dog-leg right and forward, takes the car to its quoted maximum speed of 128 m.p.h. at just under 6,000 r.p.m., although the car I was driving seemed quite happy to go right up to the speed at which the rev. limiter came into operation — some 135 m.p.h., if my calculations are correct. The gear change linkage must have been modified from the early GTV's, for the change is now very pleasant particularly the downward shift from fifth to fourth, often so unsatisfactory with the dog-leg top gear configuration. Reverse, however, was sometimes very difficult to engage.

Acceleration matches the top speed and whisked the car past other traffic in an effortless manner. As previously mentioned, bottom gear feels slightly low, but this is no disadvantage as the engine pulls very well throughout a wide speed range — it is only those who love the snarl of the exhaust as the engine pulls hard in the upper speed ranges who will be upset!

The ride is firm, as one would expect, but the suspension and seating have been arranged so well that no road harshness is transmitted to the passengers and there is no body resonance at any road or engine speed. The car handles well, tending to understeer initially to encourage the driver to apply a little throttle to balance the car. Power induced tail slides are easily caught, but are not easy to achieve, as the P6 Pirelli tyres are extremely difficult to unstick under normal conditions and provide the car with rail-like cornering abilities in the dry. It was only on tree-shaded, dew-damp roads that I was able to make the tail break away.

Thus the 2.5 GTV 6 is a delightfully Italian Italian car — a worthy vehicle to carry the Alfa badge. A revamped 2000 GTV will continue alongside the larger engined car with the same styling changes (apart from the humped bonnet) and interior improvements, but with the same mechanical set up as the current 2000 GTV. Both cars are due in the UK in May or June next year, and the forecast prices are £10,200 for the GTV 6 and £9,300 for the 2000. — P.H.J.W.

Alfa Romeo GTV6 2.5

Some day you'll be asked to define the term "sports car"... When that time comes, just show 'em this Alfa Romeo.

• There are certain fundamental truths in this business. Chiseled right up near the top on the old automotive Rosetta stone is one that's held up for as long as anybody can remember: *Alfa Romeos are fun to drive.* And there's a sub-truth: *Alfa Romeo coupes are REALLY fun to drive*, especially when they have 154-horsepower six-cylinder engines and five-speed manual gearboxes. The Alfa Romeo GTV6 is so much fun that we would all lose our driving licenses within a month if we owned one. The engine noise, all by itself, is enough to push any one of us to hitherto unscaled heights of irresponsibility.

Leave the office, wander out into the parking lot with the keys in your hand. Where is it? Ah, there. It cannot be missed. The only thing it looks like is another Alfa Romeo GTV. Black, with dished alloy wheels, a neatly integrated air dam, and a *non sequitur* matte-gray insert in the middle of the hood's wedge-shaped bulge (there to accommodate the V-6 engine). Insert a key that looks like a metal tongue depressor, unlock the door, and flop into the traditional Italian "Ciao, Sophia," driving position. In order to reach the top of the steering-wheel rim or the dashboard controls, you must put the seat's backrest into the full-upright position. This brings your head into contact with the headliner. So you crank the backrest down a degree at a time, trying to find the perfect compromise among headroom, splayed-out legs, attitude of command, and the necessary leverage to turn the wood-rimmed steering wheel. Forget it, there is no such compromise. In this car, you'll always be giving up something in your final selection of a driving position, but after a half-hour on the road you won't care.

First gear is too short, and the shift linkage is vague and suffers from too much detent, but pretty soon you won't care about that either. What you have here is a time machine. Return with us now to those thrilling days of decent power-to-weight ratios and right-now throttle response. You turn the key, poke it into first gear, and . . . it goes "graunch," and jumps back at you when you start to let the clutch out. Clutch in again, pull the lever across the overly strong detent, and punch the sonuvabitch into first *con brio.* This time it's well and truly in gear. Let the clutch out. It idles away, smooth as glass. Give it a little throttle. Nothing happens. Give it some more throttle. It stops. Aha! That was the brake you had your foot on. The brake and throttle pedals are so close together—for ease of heel-and-toe operation—that you will leap in and put your big old Frye boot on the wrong one from time to time. Not to worry, it never happens when you're on the last lap of the Targa Florio, taking Alberto Ascari on the outside as you head for the checker. (You'll find yourself doing a lot of that in this car.)

Once you've sorted the pedals out, you're in business. You hammer it in first gear, and you think it's going to snort and paw the air. In a little more than two and a half seconds you've hit 30-plus miles per hour, used up first, and you're going for second. It takes a total of 8.3 seconds to get to 60 miles per hour, and you're still accelerating smartly, but you're in third gear and the rate of acceleration is falling off. It will take you almost fifteen seconds to get from 60 to 100. Top speed is about 130, but that really does take a while. It's terra incognita up there anyway. This is a car that's most lovable between 45 and 75 on nasty little winding roads. Not only does it go fast, it always goes where it's pointed. Better still, it stops as though it had caught an arrester cable. Masochistic little minx that it is, the GTV6 fairly thrums with pleasure when thrashed. Goes like hell, sticks like glue, puts the roses back into the sallowest cheeks. You could spend a lot more than this Alfa's $17,000 purchase price on a shrink and get nowhere near the same amount of therapeutic uplift for your money. The same is unfortunately true of the $24,000 you'd spend for a Porsche 924 Turbo.

It would be silly to suggest that $17,000 isn't a lot of money, but this car is a bargain, whether you compare it with other cars in the same general performance range or with cars that require roughly the same kind of mortgage. At the same time, it is not just anybody's $17,000 sports coupe. It is a true sports car for a guy who loves to drive hard and fast, a guy who isn't afraid to do a little of his own maintenance and repair work. For, truth to tell, the Alfa Romeo GTV6 is not perfect. No Alfa Romeo has ever stayed in our motor pool for more than a week without something minor and irritating going wrong with it—most often something electrical. On this one, the wipers blew their fuse with boring regularity. The local Alfa people took a look and announced that they had fixed a short. We gave the old wipers a test on a bright, sunny day and they were fine, worked like a charm. But it snowed that night, alas, and next morning when we turned them on they didn't work. This is typical of Alfas and, we hasten to add, detracts very little from their old-fashioned charm. If you want the day-in, day-out, never-fail reliability of an American sedan, buy an American sedan—or maybe something Japanese. On the other hand, if you want to grin like a fool every time you go screaming through the uphill left-hander just beyond the bridge, buy an Alfa exactly like our test car. Just make

sure that you have a full set of tools and an up-to-date membership in the Alfa Romeo Owners' Club. The car may do a couple of things wrong, but it does so many things so right that the true enthusiast driver will not be disappointed.

Handling and roadability are definitely the car's strong suits. We often felt that this car's predecessor, the Sprint Veloce, had everything it needed, save horsepower; that indeed one could hardly tell whether the four-cylinder U.S. version was a good road car or not because it ran out of power so quickly. All that has changed. The new six-cylinder engine provides all the urge one could wish for, and one very quickly finds himself probing the outer edges of the Alfa's capability envelope. Its behavior falls somewhere between neutral and mild understeer, and there's always a reassuring feeling that the wheels are at right angles to the road surface with no intention whatsoever of letting go. Everything the front wheels feel, you feel; not in any jarring, unpleasant way, but as useful information, sort of subliminally fed from steering-wheel rim to finger tips. You can actually feel the front tires' footprints interacting with the road surface, and it's nice.

The chassis layout is an interesting one, predating the Porsche 928's in a couple of particulars. The engine is at the front, the two-plate clutch and the transmission at the rear, as in the 928, but the Alfa's clutch linkage feels more positive than the Porsche's—there's none of the rubbery, reboundy sensation that one experiences with the 928. On the other hand, the Alfa's shift linkage leaves a bit to be desired, being vague and elastic and requiring a certain amount of driver concentration if consistently crisp shifts are to be achieved. The front suspension consists of unequal-length control arms with torsion bars and an anti-sway bar. The rear suspension is de Dion, with two diago-

Technical Highlights

•Italian auto engineers have never been known for their eagerness to march down well-trodden design paths. A look beneath the taut styling of the Alfa GTV will tell you that it's no exception.

The all-aluminium V-6 was tooled up in 1979 to power the Alfa 6 sedan, a car originally aimed at America, then canceled from our model line-up. This engine is a jewel with 60 degrees between banks, an ultra-short stroke, and large, direct ports servicing hemispherical combustion chambers. One belt-driven camshaft per bank operates the intake valves directly, and the exhaust valves, through a short stiff, pushrod-and-rocker-arm arrangement. Bosch L-Jetronic fuel injection feeds the V-6, producing a wide torque band and better than one horsepower per cubic inch.

The light, 2.5-liter V-6 sends power down a short two-piece driveshaft to the rear half of the powertrain: a two-plate clutch, a five-speed transmission, a final-drive differential, and in-board disc brakes, all carried in (or on) one webbed-aluminium assembly.

The GTV uses torsion bars with unequal-length control arms in its front suspension and coil springs with a de Dion axle in back. The rear layout offers low unsprung weight but lacks left-from-right independence. Both wheels are carried on a rigid, triangular assembly made up of two diagonal trailing members and the de Dion tube. Lateral location comes from a transverse Watt linkage. Since the transaxle is chassis-mounted, torque is transmitted to the wheels via half-shafts. Alfa's token acknowledgement of convention in the GTV consists of normal anti-sway bars, front and rear. – *Don Sherman*

Front suspension

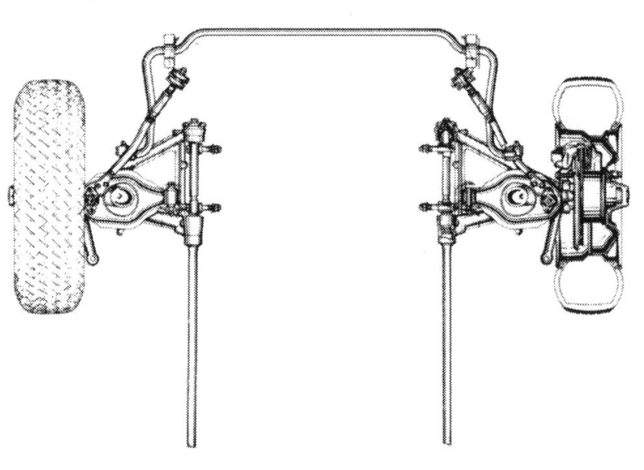

Rear suspension

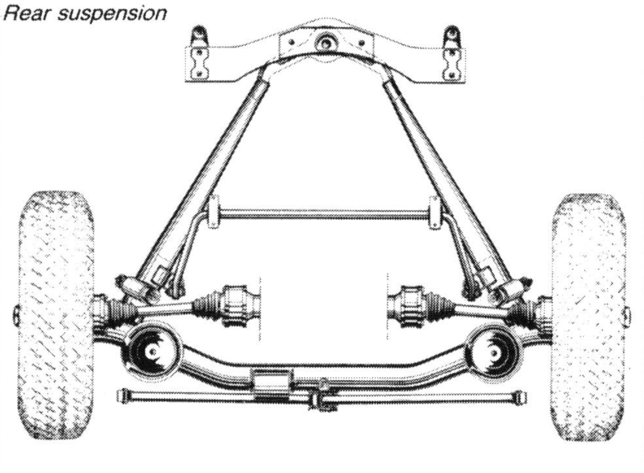

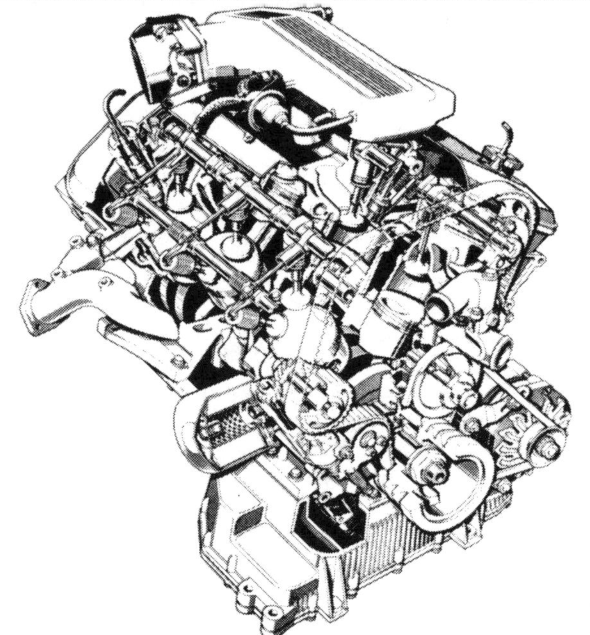

2.5-liter V-6 engine

Transaxle

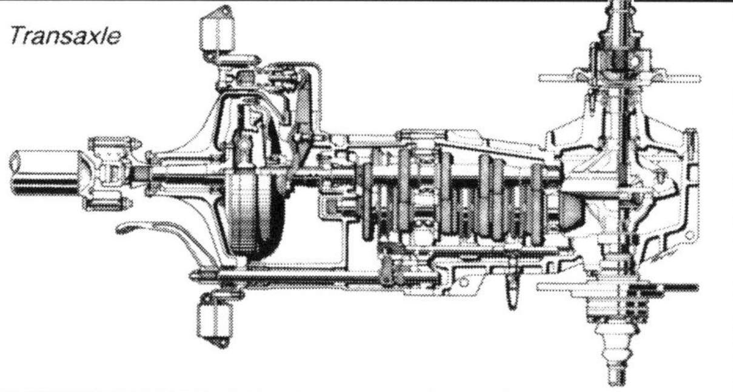

nal trailing arms integral with the axle tube, a transverse Watt linkage, coil springs, and an anti-sway bar. The fifteen-inch Campagnolo wheels are cast magnesium alloy with six-inch rims, carrying Pirelli 190/60HR-15 P6 steel-belted radials. The layout is completed with disc brakes at all four corners, the fronts being ventilated.

If you were looking for one word to sum up this new Alfa Romeo sports machine, "balance" would spring to your lips. Everything about the car shows a balanced, sensitive approach to the needs of a demanding enthusiast driver. It has ample power to start with. It exudes sporting character, yet it is mellow and quiet on the four-lane. The 2.5-liter V-6 engine from the underwhelming Alfa 6 sedan—which received a mixed reception from the motoring press, at best—really comes into its own when plugged into the GTV coupe and brought to life with Bosch L-Jetronic fuel injection. Everything that needed beefing up to accommodate the larger engine is beefed up—a refreshing change from all those contemporary cars that find themselves progressively un-beefed with each new round of cost-cutting—and the brakes, the suspension, and the powerplant all blend together in a truly balanced whole. The Alfa Romeo GTV6 is not for everyone, but happily, Alfa Romeo only has a tiny handful of fifteen hundred of these cars to distribute among America's keen types this year. We hope every one of them will find its way into the hands of people who'll enjoy it as much as we enjoyed our test car.

—David E. Davis, Jr.

COUNTERPOINT

Last week I drove the GTV6 to a function at the Detroit Golf Club – one of those fancy-schmancy places where the help is better than you, and where a nice young man idles by the front door, waiting to park your car so you don't have as far to stumble in your spike heels. "Careful with the car, buddy," I say, handing over the Alfa's keys. "We've had better cars around here than this, lady, " says the nice young man.

(Now, little does the nice young man know that in the very bowels of this unobtrusive-looking little car, behind the tacky plastic dash with its hodgepodge of misplaced instrumentation and goofy temperature controls and air vents, beyond the uncomfortable set of the steering wheel, and ultimately connected to the very touchy shift lever, simmers a tach-spinning, heart-thumping bear of an engine. I am not insulted.)

A doorman assists me into the vestibule. Behind me, I hear, "errrt, VrrrroooOOOMM, errrt, VrrrroooOOOMM, errrt, Vrrrr-rooooo OOOOOOOOMMMMMMMMM-MMMM..."

-Jean Lindamood

I've often wondered what it is that the Porsche 924 lacks, and this Alfa showed me: an engine with soul. The V-6 gives the GTV a new, lusty persona, but the power increase is only part of the story.

Aesthetics, I realized, count for an awful lot as well. And the GTV's new engine gives good aesthetics. Its idle is lumpy and mean. It burbles seductively at low revs. It winds up effortlessly. And when you lean hard on the throttle, it crescendos into a rich raspy, biting yowl–ah, Formula 1 sounds. It reminded me a lot of the Porsche 911's terrific flat six. Thanks to the new V-6, the Alfa fairly tingles with boy-racer tension, and that's what cars like this are all about - right?

The 924's four-cylinder engine, on the other hand, sounds tinny, nasal, and strained. You almost feel bad using it hard. But all that wonderfulness throbbing away on the other side of the GTV's firewall made me want to wave my arms around and drive like crazy. What a difference. 'Atsa nice.

-Rich Ceppos

I've got roughly ten years of Alfa's under my belt by now. Invariably, they've always been nice, puttering around cars – zesty coupes, classy sedans, sunny-Sunday roadsters, but never really quick. I thought I had run the gamut, but boy, was I wrong. The new GTV admittedly looks a bit dowdy, but I soon learned this is merely part of a dastardly disguise aimed at maximising the humiliation of Datsun 280-ZX, Porsche 924, and Mazda RX 7 (count me out) owners. They very quickly find out this new Alfa is anything but slow. For example, I logged the New York-Ann Arbor run in a painless nine hours flat (70 mph average) without super heterodyne assistance. To beat that, I piloted the Alfa at the track, where it set the curve in practically every test: 70 to) mph in 182 feet, 0 to 60 in the low eights, and a nice round 130 mph if you've got four or five miles to let it rip. What more could you ask for? Final calculations showed the GTV had produced 24 mpg through these heavy-throttle exercises. So, Alfa, give me seating-position and shift linkage improvements and I'll be back to work out a lease program that will hold me over the next ten years.

-Don Sherman

Vehicle type: front-engine, rear-wheel-drive, 4-passenger, 3-door sedan

Price as tested: $17,000 (estimated).

Sound system: Blaupunkt CR-3001 AM/FM-stereo radio/cassette with equalizer/booster, 2 speakers

ENGINE

Type:	V-6, aluminum block and heads
Bore x stroke	3.46 x 2.69 in, 88 x 68mm
Displacement	152 cu in, 2490cc
Fuel system	Bosch L-Jetronic fuel injection
Power (SAE net)	154 bhp @ 5500 rpm
Torque (SAE net)	152 lbs-ft @ 3200 rpm
Redline	6300 rpm

DRIVETRAIN

Transmission	5-speed
Final-drive ratio	4.10:1

Gear	Ratio	Mph/1000 rpm	Max. test speed
I	3.50	4.9	31 mph (6300 rpm)
II	1.96	8.7	55 mph (6300 rpm)
III	1.34	12.8	81 mph (6300 rpm)
IV	1.03	16.7	105 mph (6300 rpm)
V	0.78	22.0	130 mph (5900 rpm)

DIMENSIONS AND CAPACITIES

Wheelbase	94.5 in
Track, F/R	54.1/53.8 in
Length	167.7 in
Width	65.5 in
Height	52.4 in
Curb weight	2840 lbs
Weight distribution, F/R	50.7/49.3%
Fuel capacity	20.0 gal

CHASSIS/BODY

Type	unit construction
Body material	welded steel stampings

INTERIOR

SAE volume, front seat	45 cu ft
rear seat	31 cu ft
trunk space	7 cu ft
Front seats	bucket
Recliner type	infinitely adjustable
General comfort	good
Fore-and-aft support	good
Lateral support	good

SUSPENSION

F: ind, unequal-length control arms, torsion bars, anti-sway bar

R: de Dion, transverse tube integral with two diagonal trailing arms, transverse Watt linkage, coil springs, anti-sway bar

STEERING

Type	rack-and-pinion
Turning circle curb-to-curb	32.8 ft

BRAKES

F:	10.5 x 0.9-in vented disc
R:	9.8 x 0.4-in disc

WHEELS AND TIRES

Wheel size	6.0 x 15 in
Tire make and size	Pirelli Cinturato P6, 195/60HR-15

Car and Driver Test Results

ACCELERATION

	Seconds
Zero to 30 mph	2.6
40 mph	4.2
50 mph	5.8
60 mph	8.3
70 mph	10.8
80 mph	14.1
90 mph	18.2
100 mph	23.2
110 mph	34.0
120 mph	49.3
Top-gear passing time, 30–50 mph	9.6
50–70 mph	10.6
Standing ¼-mile	16.2 sec @ 85 mph
Top speed	130 mph

BRAKING

70–0 mph @ impending lockup	182 ft
Modulation	excellent
Fade	none
Front-rear balance	excellent

HANDLING

Roadholding, 282-ft-dia skidpad	0.75 g
Understeer	minimal

COAST-DOWN MEASUREMENTS

Road horsepower @ 50 mph	13.0 hp
Friction and tire losses @ 50 mph	6.0 hp
Aerodynamic drag @ 50 mph	7.0 hp

FUEL ECONOMY

EPA city driving	**17 mpg**
EPA highway driving	29 mpg
EPA combined driving	21 mpg
C/D observed fuel economy	**24 mpg**

INTERIOR SOUND LEVEL

Idle	64 dBA
Full-throttle acceleration	83 dBA
70-mph cruising	76 dBA
70-mph coasting	74 dBA

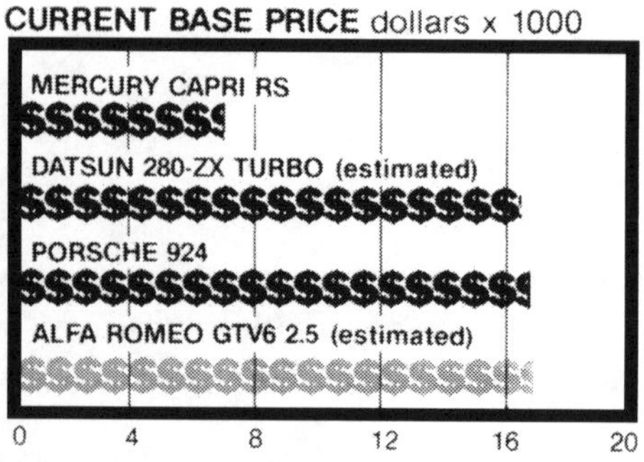

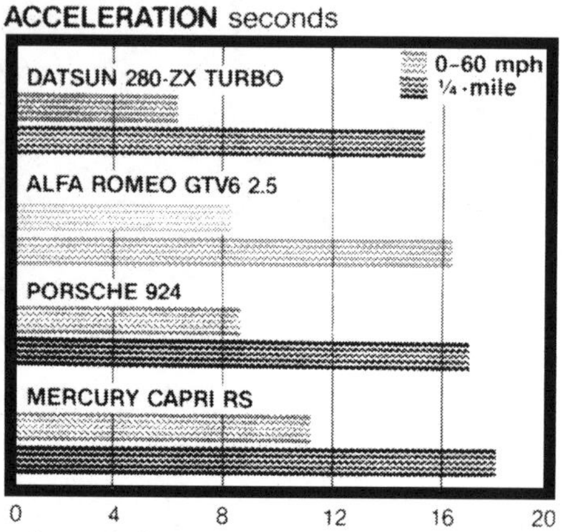

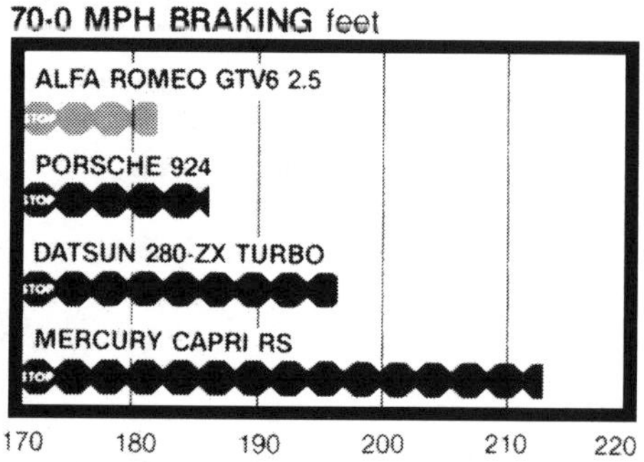

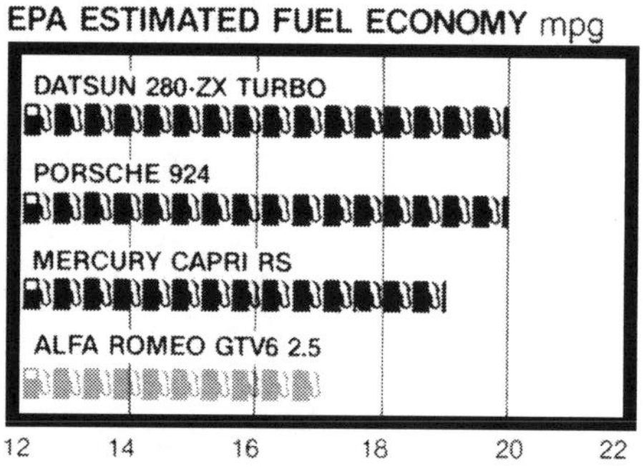

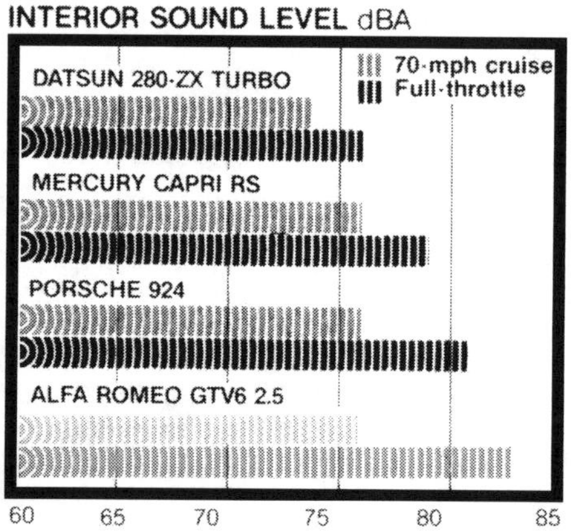

Alfa Romeo's GTV6

some further thoughts

STANDARD HOUSE is quite an Alfa Romeo-orientated place and we reported favourably last year on the Alfa Romeo 6 saloon and have only recently enthused over the Alfa GTV 6 2.5 coupé with the clever all-aluminium / magnesium 60 deg. V6 2½-litre engine using a combination of o.h.-camshaft and short-push-rod valve-gear in the former GTV two-door body shell. A further full report is not called for, but having now driven the latter model I can fill in with some comments on using it for an enjoyable thirteen days in this country.

WITH its speed bulge in the centre of the bonnet, to provide clearance for the large air-cleaner, and its low profile Pirelli tyres, the GTV6 looks extremely aggressive, but rather less elegant than its 2-litre predecessor (which is still available). Inside, creature comforts are well finished, and the driver's seat is provided with height adjustment as well as the usual fore-and-aft and rake. The height adjusting lever can be seen low down alongside the seat. Clutch travel is rather long, but there is a comfortable rest for the left foot. The rim of the steering wheel is thick, and in warm weather, the hands can slip on the polished wood surface. A surprising omission on a car of such undoubted quality is the fitting of electric windows — much more useful than the electric adjustment of the single exterior mirror.

On the run out of London through the usual congestion up to the M40 the transmission proved the unfortunate factor. The clutch needs a long foot-movement, unsuited to one who has short legs and likes to sit at anything but an arms-stretch position, yet it bites only at the end of this travel. Some hasty seat re-adjustment was necessary before jerky novice starts were eliminated, and this did not overcome a gear-change encompassing strong spring-loading to the centre of the 5-speed gate, a very baulky 1st gear, and sticky movements into and out of the lower ratios, of which bottom and second are too low. On the other hand, once clear of the traffic the flexibility of the new V6 engine in fifth speed, and the pleasant and quick shift back from fifth into normal top gear made up, to some extent, for the very hard work the Alfa gearbox imposes on the driver in traffic.

Another early impression was the eager exhaust-note as the engine was opened up, even more sporting than that of an Alfetta. Once past Oxford and all the old charm and appeal of Alfa Romeo motoring became quickly apparent. I found, as the miles accumulated, that not only was I covering the ground very quickly but that I was undoubtedly enjoying myself and that I was passing in places where even the rapidly-accelerating Rover 3500 would have made me hold back. Such is the security built into precision Alfa 6 road-holding! The suspension, firm at low speeds, soaks up bad roads in the Alfa tradition.

Apart from the pleasure of driving this 130 m.p.h., 0-60 m.p.h. in 8½ sec., GTV 6, there is the allure of that splendid Bosch-fuel-injection 160 b.h.p. engine, De Dion transaxle rear suspension, torsion-bar front suspension, Pirelli P6 tyres on magnesium wheels and disc brakes all round, the rear pair inboard mounted. Steering is manual, which accounts for much effort when

Continued on page 57

THE all alloy V-6 engine has been described fully in earlier issues of MOTOR SPORT. It provides very smooth power up to its maximum of 6,000 r.p.m., and this is accompanied by an exciting, very "Italian", exhaust note. The light alloy Campagnolo road wheels are shod with 60 profile P6 tyres of 195 section, and the forward portions of the rear windows wind down.

Auto TEST

Alfa Romeo GTV6

Refined, fast, frugal, but . . . .

Metal trim is black finished, GTV6 also features revised bumpering and front spoiler, plus angular bonnet bulge, with inset moulded "styling panel". Cornering roll is moderate, with only gentle understeer at the limit of adhesion

IN ADDITION to turning out probably the best handling small car of the decade from their Naples factory in the shape of the Alfasud, Alfa Romeo "Nord" have consistently produced intriguing designs in their larger cars. Advanced even by today's standards, the Alfetta GTV clearly fell into this category when it first appeared in 2-litre form in May 1976; a shapely coupé following Alfa's then quite recently introduced De Dion rear axle, plus rear mounted transaxle and clutch concept, with the attendant theoretical benefits in handling balance traction and ride – certainly a chassis that would readily accept more power without fundamental change.

In fact the GTV6 (introduced last December with the face-lifted 2-litre) incorporates many detail modifications in addition to the powerplant which is of course Alfa's new very oversquare (88 x 68.3mm bore and stroke) all alloy, wet linered, 60 deg V6. This has an interesting head design where single overhead camshafts work the inlet valves directly, and the exhausts via a cross-over pushrod arrangement. Thus Alfa achieve a hemispherical combustion chamber with only one belt-driven camshaft per bank. With Bosch L-Jetronic fuel injection (dictated by US emission laws), rather than the six Solex or Dellorto carburettors found on the Alfa 6 saloon, power is quoted as 160 bhp (DIN) at 5,600 rpm (rather than at 6,000 rpm as was originally stated) with 157 lb. ft. torque at 4,000 rpm.

Transmission modifications are extensive, with stronger propshaft couplings, a twin plate clutch and a rear transaxle containing modified gears and larger bearings, also lower than 2-litre first and second gear ratios – a quite illogical move for which Alfa has yet to provide a satisfactory explanation. Fifth gear is raised giving the GTV6 overall gearing of 22 mph per 1,000 rpm.

As before, suspension is by double wishbones in front, the solid de Dion rear axle being

Alfa Romeo GTV6: Introduced December 1980. Uses Bosch L-Jetronic fuel injected version of the 88 × 68.3mm all-aluminium V6, first seen in the Alfa 6 saloon in six carburettor form. GTV6 features spring and damper rates revised upwards from GTV 2-litre, front discs, ultra low profile tyres, new moulded front and rear lower body sections and side sills, plus a distinctive power bulge. UK sales started earlier this month.

PRODUCED BY:
Alfa Romeo S.p.A.,
Stabilimento de Arese,
Milan 93391
Italy

SOLD IN THE UK BY:
Alfa Romeo (GB) Ltd,
Edgware Road,
London NW2 6LX

located longitudinally by trailing arms and laterally by a Watts linkage. Front torsion bar and rear coil spring rates are raised slightly as are damping rates. Heavier anti-roll bars are fitted. The front discs are now ventilated.

Outside the GTV6 is easily recognisable by its Campagnolo alloy wheels fitted with ultra low profile 195/60HR-15 Pirelli P6 tyres, the distinctive bonnet bulge, also by the new moulded bumpering, side sills and black finish metal trim which is also common to the 2-litre. The latest cars look cleaner yet Alfa say no measureable improvement in aerodynamic efficiency results, leaving the GTV6 at the previously quoted drag co-efficient figure of 0.39.

Inside it came as a real disappointment to find that Alfa have retained their peculiar single isolated speedometer and centre-mounted main instrument binnacle arrangement, instead of incorporating the utterly sensible and conventional closely grouped instrument layout found on left hand drive cars.

Installing the V6 engine could easily have produced a significantly more forward weight bias. As it is, the extra frontal weight is balanced by the heavier transmission and fuel tankage. The test GTV6 tipped the scales at 24.1 cwt (Autotest GTV 2-litre 21.6 cwt) distributed within a fraction of the same ratio at 51/49 front to rear.

Performance
Silky

As we have come to expect in Bosch fuel injected engines, start-up was immediate. Driveability during an unusually long warm up period was excellent – certainly there was no need to use the hand throttle to prevent stalling, even though the tickover on this test car seemed to be set low (500 rpm) resulting in a rather uneven idle. Those expecting the Alfa V6 to provide something extra in the way of mid-range performance may be a little disappointed. Those looking for free revving smoothness, refinement and flexibility, on the other hand, will be overjoyed by the manner in which the Alfa performs. This was aptly illustrated during a 700 mile trip well into Germany *and back to London* within 24 hours where the car would cruise quite effortlessly at between 100 and 120 mph and when asked rev smoothly round to the 6,300 rpm red line and rev limiter (6,400 rpm on the test car) in the gears.

At these crankshaft speeds valve gear noise becomes quite marked, but the engine still remains vibration free. Indeed, the lack of any significant vibration period throughout the rev range adds enormously to the feeling of sewing machine-like mechanical refinement.

As we have commented before, gearchange movements in

Campagnolo cast alloy wheels and Pirelli P6 tyres are standard fitments. Dark grey plastic mouldings cap rear bumper, and mask rear quarter ventilation outlets. Bright metal guttering is capped with flexible plastic strip

the transaxled Alfas leave something to be desired, the GTV6's being especially sticky and heavy between the lower gears, if otherwise free and precise enough. Engagement of first is always accompanied by a crunch, or simply baulks if a more leisurely approach is tried, a seemingly inevitable result of the rotating inertia of flywheel and propellor shaft having to be overcome by the synchromesh. Second gear also rarely goes in without a clonk, the first to second change being made awkward by the engine rpm dying away slowly, a clutch that takes up near the end of its travel, and strong spring loading of the gear-lever towards the third/fourth plane. Furthermore first and second gears are irritatingly low offering maxima of only 31, and 56 mph. With third going all the way to 81 mph one finds oneself having to make the awkward downchange from third to second against that spring pressure more frequently than one would like during brisk cross country drives. Reverse – opposite fifth, towards the driver and back, proved particularly difficult to engage, as if there was some hangup in the linkage.

With two shifts before 60 mph, acceleration from rest to that speed suffers a little, however when performance testing the low first gear did enable wheel spinning starts to be easily made. The GTV6 went most respectably to 30, 60, 100 and 110 mph in 2.8, 8.8, 24.3 and 34.6 sec. Some driveline snatch prevented us taking in the gear figures from 10 mph in fifth, otherwise the Alfa would accept full throttle from this speed pulling smoothly away in the lower gears in one long flow and without any sensation of "camminess", or quite the immediacy of larger engined, yet similarly geared, competitors like the 2.8 Injection Capri and TVR Tasmin which for example would do 50-70 mph in top in 8.2 and 8.7 sec respectively against 9.7 sec for the GTV6.

The pleasing part is that the engine shows no sign of running out of breath way above the quoted power peak as was shown when it came to maximum speed testing, the Alfa exceeding the factory claim of 128 mph by recording a best one-way speed of 132 mph and a mean maximum of 130 mph with the engine revving at 5,900 rpm (300 rpm above peak power). In spite of this apparent undergearing, mechanical refinement is such that one does not feel the car necessarily needs a higher top, though it would obviously pull a slightly taller one.

Economy
Potentially excellent

It came as a pleasant surprise to find that on long fast runs the GTV6 was capable of better than 25 mpg. Only by driving the car very hard could we bring interim brim figures beneath 22 mpg, an interestingly small variation. The overall figure of 23.2 mpg falls neatly between these extremes, such figures suggesting the GTV6 should manage 25 and 27 mpg in more normal use. The tank holds 16.5 gallons. London-Cologne (330 miles) is easily achieved without refuelling, though the fuel reserve light is likely to have started blinking near journey's end, leaving a comfortable reserve of 3 gallons. Little time is lost at the filling station since the Alfa's tank brims immediately to the top – most pleasing. In our experience relatively complex all-aluminium engines do tend to use a little oil, the Alfa 6 unit being no exception, needing a pint every 800 miles.

Noise
Best in class?

The GTV6 is blessedly free from any body boom periods or engine vibration – any "death ray" effect, often unnoticed at first, but encountered in some cars only after long periods at high cruising speeds. Apart from the period of fussiness right at the top end the engine spins beautifully up to the rev limiter with a purposeful yet always muted exhaust note. This cuts back to a murmer on light throttle openings. For an ultra low profile tyre shod car the Alfa is also particularly well insulated from road noise, most poor surfaces taken at moderate speeds provoking a distant rumble, only the worst road joints and potholes causing significant thump.

Coarse concrete surfaces do produce noticeable tyre roar – quite loud on the very worst rain grooved sections of *autoroute* in Belgium, yet the occupants are better isolated from this source of constant stress than in most similarly shod cars, to the point where it does not become an eventual irritant. Double door sealing and the steeply raked screen can only contribute to the GTV6's low wind noise levels. At cruising speeds over 70 mph there is some from around the A-posts, but in the 100-110 mph bracket it has not increased much, thus adding to the feeling of relaxed refinement at speed. It seems that owners will simply have to get used to the frequent clonks and crunches that accompany first and second gear selection, and the gear rattle (quite commonly found in other cars) emanating from the transaxle at tickover.

Road behaviour
Effortless

Even with the increased tyre width Alfa have not fitted power steering, nor, thankfully, have they changed the previous GTV's near-ideal gearing. The 3.6 turns of the wheel cover a 34ft mean turning circle. Though parking efforts are high the steering lightens to the ideal degree once on the move. At high speeds one notices there is some free movement about the straight ahead, so that deviating the car from the straight requires more than the initially expected degree of lock – as if one has to take up some rubberiness. Thereafter steering loads build beautifully in relation to cornering speeds, the driver always sensing accurately what the front wheels are doing. At any reasonable cruising speed straight line stability is excellent with good resistance to crosswinds, the car running true with little attention.

That said, we did notice that when running at very high speeds, say 115 mph and above, the car became much more easily deflected from its track, the steering lightening, and the front end becoming quite eerily vague, a characteristic we did not encounter on the factory-prepared launch cars which ran arrowlike right up to their maximum, another quirk being a tendency (noticed in previous GTVs) for bumpy roads to jog the front offline while running straight.

Once the suspension is loaded up in corners, the GTV6 sticks to the road beautifully with little roll and without being at all upset by mid corner surface changes or bumps. For the most part the driver feels wonderfully in contact with the road, being able to place the car accurately, and easily sense the limit of adhesion as it approaches.

In extremis the extra power of the GTV6, improved chassis and extra grip, combine to make it a much more neutrally balanced car than the first 2-litre GTVs. There is just the right degree of understeer to *lead* the driver into a corner. At the same time the impurturbability of the rear end and excellent brake balance permits the brakes to be used as one is turning in without upsetting the car. This and the drivers' ability to modify his understeering line by backing off the throttle (which produces only a gentle and moderate tightening of line) makes the GTV6 a very satisfying car to drive on the road or test track. Only track driving techniques and full throttle in second gear in the wet would provoke any significant rear end slide. Such manners translate into a car that covers the cross country miles fast without effort or stress, perhaps more than usually so because Alfa have also done so well on ride. At low speeds it is taut. Wheel movements are closely controlled but absorb sharp surface irregularities with an uncommon suppleness. Long undulations are soaked up beautifully, also B road surfaces, though some bumpiness is transmitted on the roughest ones, a characteristic which as we have already mentioned, one occasionally encounters at high speeds on sections of rippled motorway.

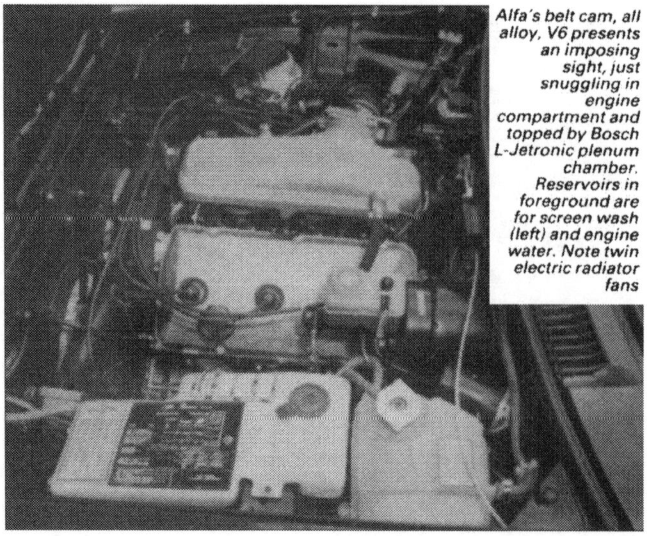

Alfa's belt cam, all alloy, V6 presents an imposing sight, just snuggling in engine compartment and topped by Bosch L-Jetronic plenum chamber. Reservoirs in foreground are for screen wash (left) and engine water. Note twin electric radiator fans

Brakes
Excellent

The brakes are yet another excellent feature of the GTV6's dynamic package; perhaps a shade light for check braking, with a 20lb load producing 0.38g retardation, but having excellent crash stop control since the pedal pressure increases markedly when braking heavily, 50lb giving 0.75g, and 90lb an excellent peak of 0.98g with the front tyres on the point of breaking traction. Thankfully the early front locking so common nowadays is not a feature of the GTV6's braking. Fade resistance is excellent, pedal pressure rising only marginally towards the end of the test (10 consecutive 0.5g stops from 84 mph). Thanks to the GTV6's weight distribution the hand brake managed an easy 0.4g stop on the flat, the long travel lever almost running out of movement then and when arresting the car on MIRA's 1-in-3 test slope. The restart was an absurdly simple matter due to the car's low first gear.

Behind the wheel
Awkward

Again we must confess to being very disappointed at finding the old style GTV facia on right-hand drive cars, with the speedometer (unusually 3 per cent slow at 60 mph) sitting in isolation in front of the driver, while the other instruments, revcounter, oil pressure, fuel contents and water temperature gauges plus the usual warning lights are placed out of immediate view in a separate centre mounted binnacle.

Nor, in spite of constant comment and Alfa GB's obvious desire to widen the GTV's appeal, have the factory seen fit to make the steering column adjustable for reach (though it is moveable vertically). Such comment is necessary because the driving position represents a bad example of the typically Italian one, where one has no choice but to drive with arms outstretched if one's legs are not to be cramped. We often found ourselves moving the seat back for motorway driving, then forward again in towns when a greater purchase was needed, for example when parking. The situation is further complicated because even on its lowest position tall drivers find they cannot wind the back rest fully forward without making head to roof lining contact. Most testers found an acceptable compromise in the end but all commented how much more appealing the car would be if these so easily curable ergonomic irritations did not exist. There are good points; the pedals are ideally placed for heel and toe operation, there is an excellent clutch footrest while the woodrim wheel not only looks good, adding a touch of class with its Alfa motifed centre horn press, but is pleasant to the touch. The seats are suitably firm and supportive providing good lumbar and quite adequate lateral support.

For a high performance coupé, all round visibility is better than average, with no serious blind spots, although the steeply raked screen pillars do inhibit

Roominess of front footwells is emphasised (thanks to rear transaxle) but also poor instrument layout, with centre mounted binnacle containing revcounter (with inset main beam and indicator warning lights), oil pressure, fuel contents and water temperature gauges, handbrake, heater fan, and charge warning lights. Only speedometer is before the driver's eyes. Wheel has polished wood rim, and centre horn press, while stalk controls on left work lights and indicators, the one on right the wipers and screen wash. Heater controls and radio are contained in centre console. Note well spaced pedals and generous sized foot rest

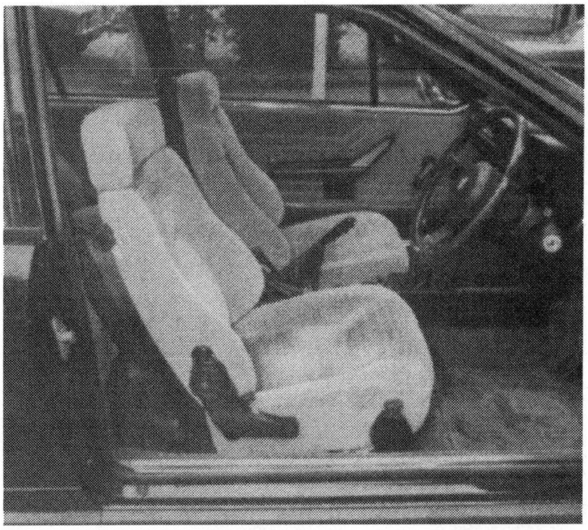

Rear seats are well shaped. Legroom is tolerable, but headroom rather limited. Front seats slide forward as they tip to make access easier.

Cockpit is spacious, and seats comfortable. Head restraints are adjustable via knob on seat back side

An extendable lever arrangement works proper parrallelogram driver's seat height adjustment. Small turnwheels in the rear quarter panels operate the wind down rear windows

Rear sill is on the high side but boot space is quite generous. Parcel shelf is inconveniently fixed to the single centre mounted tailgate gas filled supporting strut

Spare wheel is housed under floor, while jack and four piece tool kit are clipped to offside inner wing

visibility at wide angles for example when looking "across" a corner. In bright sunlight there are irritatingly clear reflections from the facia top and instrument binnacle in the screen.

The stalk controls operate tolerably well. The two on the left, working indicators and main lights respectively, are close together and of similar length therefore rather easy to confuse. Two-speed plus intermittent wipers are worked by the right hand stalk. The wiper blades are too short and leave a large unwiped section at the top of the screen irking the tall driver most.

Heating and ventilation are just adequate – certainly not in the Alfasud or Ford Capri class. Perhaps as a result of the car's much smoother lines, ram airflow (which works independently of the main system) is barely felt at speeds below the legal limit, and although much improved output can be obtained through the facia top outlets with fan assistance, the fan is noisy on the first speed and unbearably so when running flat out. There are separate directional controls for each front passenger, but not for temperature which follows Alfa's somewhat quaint water valve, plus separate air mix systems. Output is tolerable and temperature control just better than a normal water valve system – no more.

Living with the GTV

Owning a GTV6 will surely reward those who appreciate the feel of a piece of machinery, while driving or simply when opening the bonnet to reveal an engine bay dominated by two broad alloy cam covers. Daily maintenance items are straightforward to check though for convenience's sake it would be nice to see the bonnet pull and stay moved to the driver's side.

A nice touch of quality are the wind down rear quarter lights, fortunately not used so often as the maddeningly low geared window winders, which require one to reach forward in order to work them, another niggle being plastic head restraint supports that look far too springy and shoddy. The self-seeking Pioneer stereo radio/cassette (fitted as standard) worked well, receiving BBC World Service with consistent quality throughout our European trip. It features a digital wavelength display – and time display useful since the GTV6 has no clock elsewhere.

Oddments space is not particularly generous, with only a passenger's side parcel shelf, small glovebox, and large tray in front of the gearlever. There are no door pockets or seat back pouches. In contrast the boot is reasonably proportioned though one still wonders why the GTV could not have been a true hatchback. As it is the fuel tank intervenes, and a single centre mounted gas strut rear door stay is fixed to the parcel shelf, leaving a rather unflexible loadspace if bulky objects need to be carried. The spare wheel is housed in the usual place under the floor, as is the awkward to top up battery on the nearside, while the jack, wheelbrace and four piece tool kit are clipped to the offside inner wing.

Since the front seats slide forward when tipped, entry to the rear is endurable, and as long as the front seats' considerable longitudinal movement is not exploited to the full, legroom is also tolerable – for short journeys certainly. Headroom however is lacking compared with the competition.

Several hours were spent driv-

ing at night. While main beam is of good power and throw, the lights would dim markedly on dip.

We would imagine owners of this fairly complex car leaving maintenance to their local Alfa agent, but with 6,000 mile service intervals, fuel injection and breakerless ignition, it should require no more frequent service attention than is the norm.

The GTV range

There is only one model. It costs £9,495. Metallic paint is available as a no cost option.

HOW THE ALFA ROMEO GTV6 PERFORMS

Figures taken at 5,268 miles by our own staff at the Motor Industry Research Association proving ground at Nuneaton, and on the Continent

TEST CONDITIONS:
Wind: 5-15 mph
Temperature: 19 deg C (66 deg F)
Barometer: 29.5 in. Hg (1000.0 mbar)
Humidity: 60 per cent
Surface: dry asphalt and concrete
Test distance: 1,490 miles

MAXIMUM SPEEDS

Gear	mph	kph	rpm
Top (mean)	130	209	5,900
(best)	132	212	6,000
4th	107	172	6,400
3rd	81	130	6,400
2nd	56	90	6,400
1st	31	50	6,400

ACCELERATION

FROM REST

True mph	Time (sec)	Speedo mph
30	**2.8**	30
40	**4.5**	40
50	**6.3**	50
60	**8.8**	59
70	**11.7**	68
80	**14.7**	78
90	**19.4**	88
100	**24.3**	97
110	**34.6**	108

Standing ¼-mile: 16.7 sec, 84 mph
Standing km: 30.6 sec, 107 mph

IN EACH GEAR

mph	Top	4th	3rd	2nd
10-30	–	7.4	5.6	3.4
20-40	10.0	6.9	4.9	3.3
30-50	9.2	6.5	4.7	3.6
40-60	9.4	6.3	4.6	–
50-70	9.7	6.6	5.0	–
60-80	10.4	7.1	5.8	–
70-90	11.3	7.5	–	–
80-100	13.2	9.3	–	–
90-110	17.8	–	–	–

FUEL CONSUMPTION

Overall mpg:
23.2 (12.2 litres/100km)

Constant speed:

Autocar constant speed fuel consumption measurement equipment incompatible with Bosch fuel injection.

Autocar formula: Hard 20.9 mpg
Driving and conditions: Average 25.5 mpg; Gentle 30.2 mpg

Grade of fuel: Premium, 4-star (98 RM)
Fuel tank: 16.5 Imp. galls (75 litres)
Mileage recorder reads: 2.0 per cent short

Official fuel consumption figures
(ECE laboratory test conditions; not necessarily related to Autocar figures)
Urban cycle: 19.1 mpg
Steady 56 mph: 36.7 mpg
Steady 75 mph: 28.6 mpg

OIL CONSUMPTION

(SAE 20/50) 800 miles/pint

BRAKING

Fade *(from 84 mph in neutral)*
Pedal load for 0.5g stops in lb

	start/end		start/end
1	12/12	6	20/16
2	20/20	7	24/20
3	20/20	8	24/24
4	20/16	9	32/36
5	24/16	10	28/28

Response *(from 30 mph in neutral)*

Load	g	Distance
10 lb	0.20	150 ft
30 lb	0.48	63 ft
50 lb	0.75	40 ft
70 lb	0.87	35 ft
90 lb	0.98	31 ft
Handbrake	0.40	75 ft

Max. gradient: 1 in 3

CLUTCH

Pedal 30 lb; Travel 5 in.

WEIGHT

Kerb, 24.1 cwt/2,702 lb/1,227 kg
(Distribution F/R, 51.0/49.0)
Test, 27.7 cwt/3,102 lb/1,409 kg
Max. payload 750 lb/340 kg

DIMENSIONS

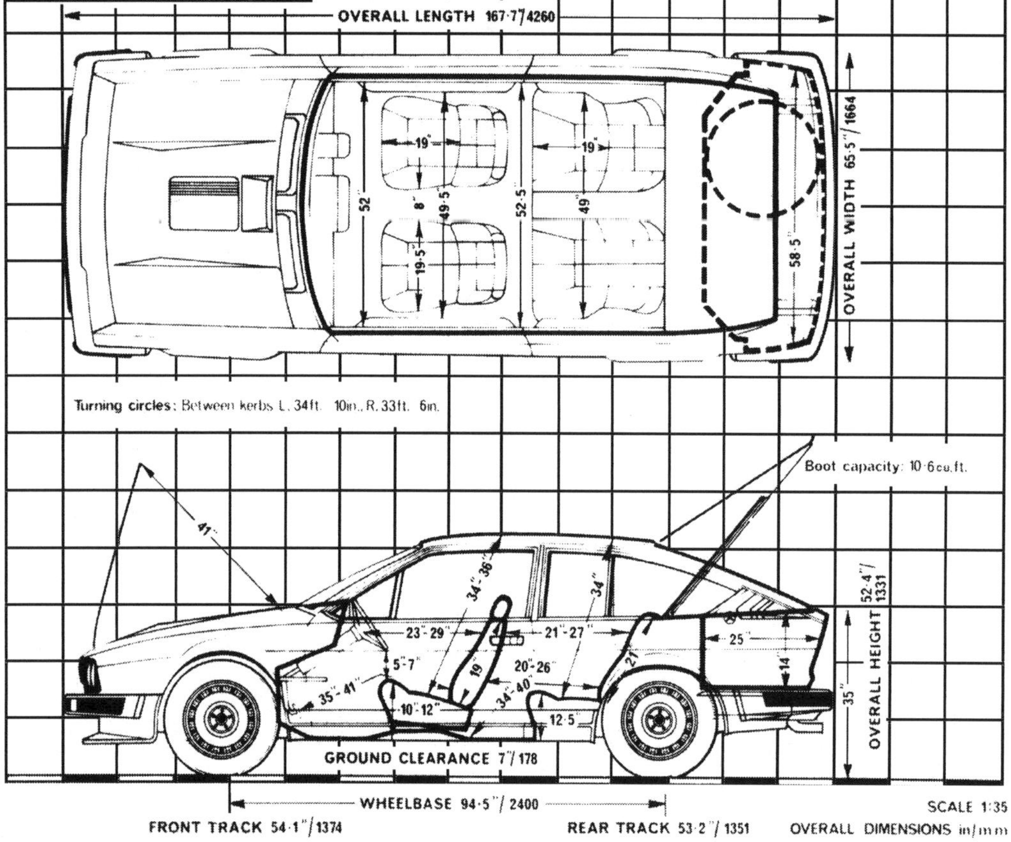

PRICES

Basic	£7,641.81
Special Car Tax	£625.15
VAT	£1,228.04
Total (in GB)	**£9,495.00**
Seat Belts	Standard
Licence	£70.00
Delivery charge (London)	£126.50
Number plates	£15.00
Total on the Road (exc. insurance)	**£9,706.50**

EXTRAS (inc. VAT)

Metallic paint	No extra charge
Manual sunroof	£350.00
Air conditioning	£500.00
TOTAL AS TESTED ON THE ROAD	**£9,706.50**
Insurance	Group not yet available

SERVICE & PARTS

Change	Interval 6,000	12,000	18,000
Engine oil	N/A	N/A	N/A
Oil filter	"	"	"
Gearbox oil	"	"	"
Spark plugs	"	"	"
Air cleaner	"	"	"

Total cost
(Assuming labour at £11.50/hour inc. VAT)

PARTS COST *(including VAT)*

Brake pads (2 wheels) – front	£16.10
Brake pads (2 wheels) – rear	£11.50
Exhaust complete	N/A
Tyre – each (typical)	£89.51
Windscreen	£102.35
Headlamp unit	£36.34
Front wing	£143.75
Rear bumper	£94.30

WARRANTY

12 months/unlimited mileage

SPECIFICATION

ENGINE

Head/block	Front, rear drive; Aluminium alloy/alloy
Cylinders	6 in 60 degV, wet liners
Main bearings	4
Cooling	Water
Fan	Twin electric
Bore, mm (in.)	88.0 (3.46)
Stroke, mm (in.)	68.3 (2.69)
Capacity, cc (in³)	2,492 (152.3)
Valve gear	Ohc with pushrod operated exhaust valves
Camshaft drive	Toothed belt
Compression ratio	9.0-to-1
Ignition	Breakerless
Injection	Bosch L-Jetronic
Max power	160 bhp (DIN) at 5,600 rpm
Max torque	157 lb ft at 4,000 rpm

TRANSMISSION

Type		5 speed all indirect transaxle
Clutch		Twin disc diaphragm
Gear	Ratio	mph/1000rpm
Top	0.780	22.0
4th	1.026	16.7
3rd	1.345	12.7
2nd	1.956	8.8
1st	3.500	4.9
Final drive gear		Hypoid bevel
Ratio		4.1 to 1

SUSPENSION

Front – location	Double wishbones
springs	Longitudinal torsion bars
dampers	Telescopic
anti-roll bar	Yes
Rear – location	De Dion tube, Watts linkage
springs	Coil
dampers	Telescopic
anti-roll bar	Yes

STEERING

Type	Rack and pinion
Power assistance	No
Wheel diameter	15.0 in.
Turns lock to lock	3.6

BRAKES

Circuits	Dual split front/rear
Front	10.5 in. dia. ventilated disc
Rear	9.8 in. dia. disc
Servo	Vacuum
Handbrake	Centre lever rear discs

WHEELS

Type	Alloy
Rim width	6 in.
Tyres – make	Pirelli
– type	P6 radial ply
– size	195/60HR-15
– pressures	F28 R28 psi (normal driving)

EQUIPMENT

Battery	12V 77Ah
Alternator	65A
Headlamps	110/230W halogen
Reversing lamp	Standard
Electric fuses	14
Screen wipers	2-speed, plus intermittent
Screen washer	Electric
Interior heater	Air blending/ water valve
Air conditioning	Extra
Interior trim	Cloth seats, cloth headlining
Floor covering	Carpet
Jack	Screw pillar
Jacking points	2 each side
Windscreen	Laminated
Underbody protection	Electrophoretic dip/ pvc underbody coating

HOW THE ALFA ROMEO GTV6 COMPARES

Alfa Romeo GTV6 — £9,495

Front engine, rear drive

Capacity
2,492 c.c.

Power
160 bhp (DIN)
at 5,600 rpm

Weight
2,702 lb/1,227 kg

Autotest
25 July 1981

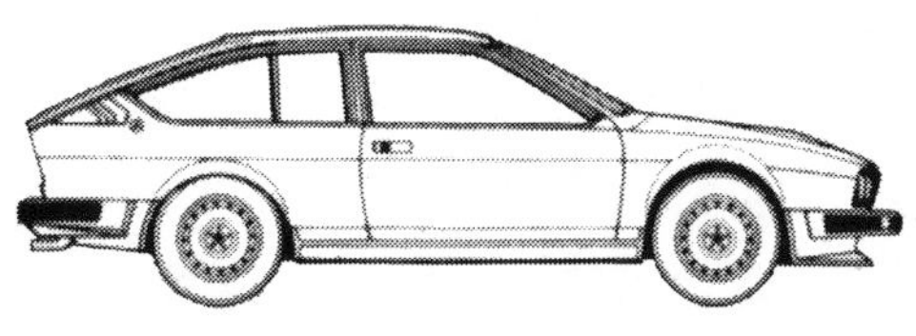

Datsun 280ZX — £9,695

Front engine, rear drive

Capacity
2,753 c.c.

Power
140 bhp (DIN)
at 5,200 rpm

Weight
2,881 lb/1,307 kg

Autotest
13 October 1979

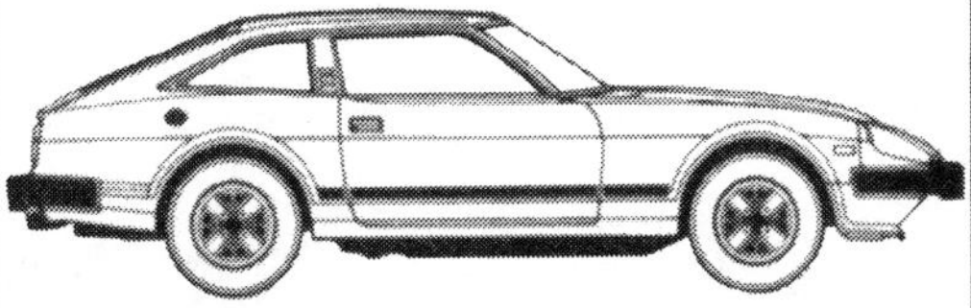

Ford Capri 2.8 Injection — £7,995

Front engine, rear drive

Capacity
2,792 c.c.

Power
160 bhp (DIN)
at 5,700 rpm

Weight
2,620 lb/1,190 kg

Autotest
20 June 1981

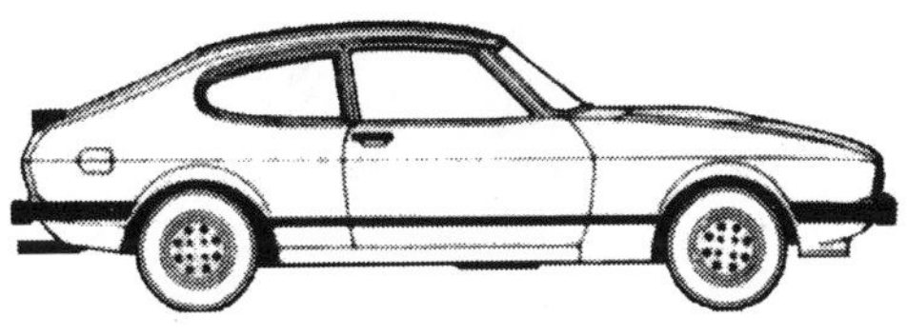

Lancia Beta HPE 2000 — £6,750

Front engine, front drive

Capacity
1,995 c.c.

Power
119 bhp (DIN)
at 5,500 rpm

Weight
2,400 lb/1,088 kg

Autotest
23 April 1977

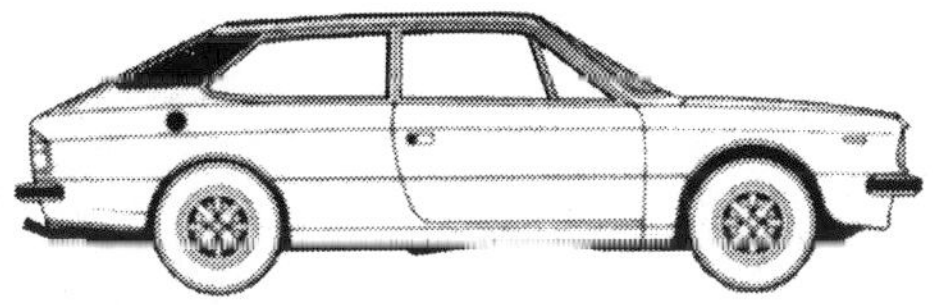

Mazda RX7 — £8,699

Front engine, rear drive

Capacity
2,292 c.c.

Power
115 bhp (DIN)
at 6,000 rpm

Weight
2,258 lb/1,024 kg

Autotest
24 November 1979

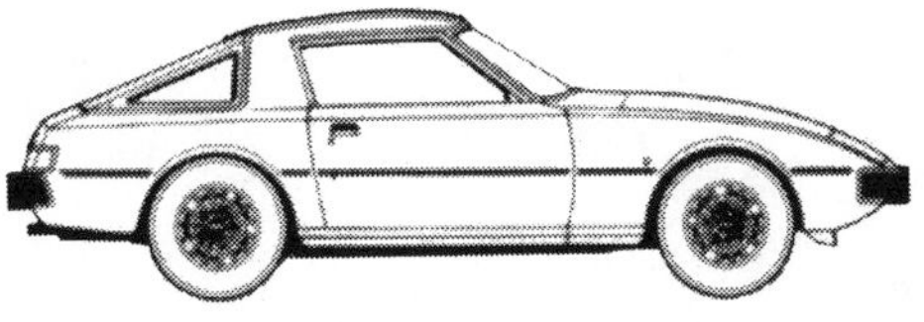

Porsche 924 — £9,103

Front engine, rear drive

Capacity
1,995 c.c.

Power
125 bhp (DIN)
at 5,800 rpm

Weight
2,450 lb/1,114 kg

Autotest
12 August 1978

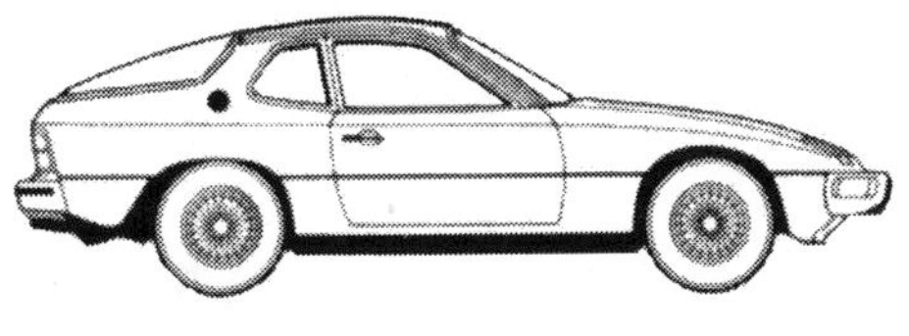

MPH & MPG

Maximum speed (mph)

Alfa Romeo GTV6	130
Ford Capri 2.8 Injection	127
Porsche 924	126
Lancia Beta HPE 2000	116
Mazda RX7	113
Datsun 280ZX	111

Acceleration 0-60 (sec)

Ford Capri 2.8 Injection	7.9
Alfa Romeo GTV6	8.8
Porsche 924	9.5
Mazda RX7	10.1
Lancia Beta HPE 2000	10.6
Datsun 280ZX	11.3

Overall mpg

Porsche 924	25.0
Alfa Romeo GTV6	23.2
Ford Capri 2.8 Injection	21.3
Lancia Beta HPE 2000	20.3
Datsun 280ZX	18.4
Mazda RX7	18.2

Other coupés worth considering might be the Audi (£7,475, 113 mph, 0-60 mph 10.? sec and 24.8 mpg) or Renault Fuego GTX (£7,127, 113 mph, 0-60 mph 10 sec, and 26 mph). Lack of test data prevents our including the considerably more expensive TVR Tasmin 2+2 (£14,904).

Having to change gear twice before reaching 60 mph mars the Alfa's initial acceleration. It still eclipses the rest on top speed, while its mechanical smoothness make it a delightfully able high speed grand tourer, a much more refined car in every respect than the Ford – a wonderfully gutsy car – or much slower and disappointingly thirsty Datsun and Mazda (both rather stodgy performers overall though the latest RX7 does have more power) and for that matter the rather noisy but pleasingly zesty Lancia. The GTV6 is much sweeter mechanically than the very aerodynamically efficient four-cylinder Porsche, and impressively, almost capable of approaching the German car on fuel economy.

ON THE ROAD

Low profile tyres give the Alfa and Capri easily the best dry road grip, while a much more sophisticated chassis means the GTV6 has the edge over the Ford on wet roads. Good traction is also the preserve of the admittedly less powerful Datsun, Mazda and Porsche, the Lancia ultimately suffering the wheel spinning limitations of front wheel drive on slippery surfaces. Best for brakes are the Alfa and Porsche, the Capri's system being very prone to front locking, and the rest averagely so.

Alfa also do well on ride in this ultra low profile tyre shod car, also road noise particularly compared with the Capri and to a lesser extent the Porsche. Neither can match the much less generously tyred Datsun, Mazda or Lancia in this respect, the Beta incidentally being an exceptionally well mannered front wheel drive machine in all respects. The Alfa earns high marks for its ultimate handling (an advanced chassis brought to life by the extra power) as do the Capri, Datsun, Porsche and Mazda. Any slight tendency for the tail to move out when lifting off in mid-corner is exceptionally easily controlled. At high cruising speeds wind noise levels are disappointingly high in the Capri, whereas the GTV6, Datsun and Porsche rate well in this area.

Number one for heating and ventilation is the Capri followed closely by the Mazda and Datsun, the Alfa's curious water valve/air blender providing barely acceptable temperature control, and fresh airflow. The Porsche and Lancia water valve systems do worse still.

SIZE & SPACE

Legroom front/rear (in.)

(seats fully back)

Ford Capri 2.8 Injection	40/36
Alfa Romeo GTV6	41/34
Datsun 280ZX	44/29
Lancia Beta HPE 2000	40/32
Porsche 924	44/23
Mazda RX7	40/25

With previous comments in mind it is galling to have to say the GTV6 still disappoints ergonomically with its awkward controls and poor wheel/seat/pedal relationship. Others – particularly the relatively roomy Capri – score heavily over the GTV6 (and to a lesser extent the Lancia) in these respects.

Of the hatchbacks the Capri and Lancia easily have the most loadable space, the Datsun's being severely restricted by the spare wheel unless Denovos are fitted. Loadable space, and rear seat legroom are very limited in the Porsche and Mazda though their opening rear windows do at least provide some sort of hatchback facility. The Alfa on the other hand only looks the part since its rear seats are fixed, the rear door merely revealing a reasonable sized boot.

VERDICT

Those with a feeling for machinery will find the Alfa appeals for its combination of smoothness, refinement, ride and handling. Its shortcomings are mainly ergonomic, nor is it a true hatchback: two areas where the Capri scores handsomely; an utterly straightforward, practical, easy to maintain high performance machine, outshining the rather sluggish Japanese contingent in all respects except ride quality, wind noise, and possibly looks when compared with the RX7.

The Lancia looks excellent value for money, and is certainly one of the sweetest handling front wheel drive sports hatchbacks. Porsche attributes are build quality, fuel economy, corrosion resistance, roadholding and aerodynamic efficiency if not mechanical or ergonomic refinement.

Above all, and more than any other car here the Alfa is a car of character. On long journeys we found its faults paling into insignificance compared with the good points, one tester summing the car up as being "a marvellous companion".

What's the best sports GT?

ROAD & TRACK SAYS:

"ALFA ROMEO GTV-6/2.5 BEST SPORTS GT... ONE OF 10 BEST CARS FOR THE 80'S."

Better than Porsche 924 Turbo, Datsun 280ZX Turbo by a significant margin!

In GT competition against Porsche 924 Turbo and Datsun 280ZX Turbo...Alfa GTV-6/2.5 won!

TOTALS:	ALFA ROMEO GTV-6/2.5	DATSUN 280ZX TURBO	PORSCHE 924 TURBO
	635	594	592

Road & Track July '81

"The first six categories of Engine, Gearbox, Steering, Brakes, Ride and Handling...do tend to get to the meat...the Alfa registered 195 points to the Datsun's 161 to the Porsche's 182."

Road & Track July '81

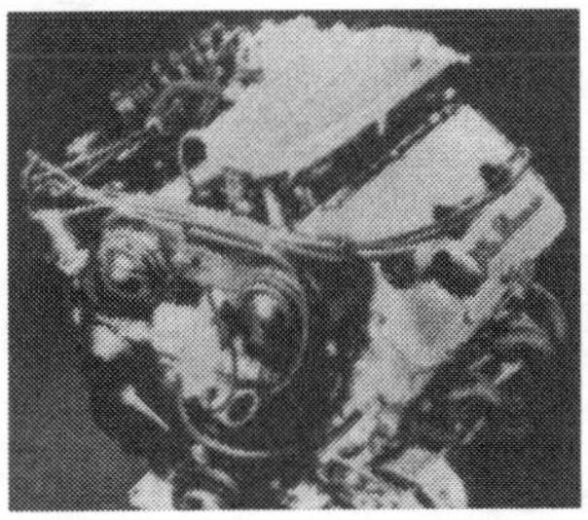

"The lovely SOHC V-6 engine...is a joy, revving and pulling smoothly through its range..."

Road & Track Aug. '81

"...We drove down some pretty vile roads at speed and were quite comfortable...the steering is light, accurate, direct...absolutely no cam effect or temperament and furthermore is extremely quiet...No twist, no wiggle, no rattle, no bad habits. Real quality." Road & Track July '81

ALFA ROMEO SAYS:

"MOST POWERFUL ALFA EVER SOLD IN AMERICA!"

Alfa Romeo

Tecnologia vincente. Da sempre.

VEE-SIX APPEAL

The new Alfa Romeo GTV6 and Ford Capri 2·8i are both powered by potent V6 engines. The Alfa costs more, but is it better? Roger Bell compares two classics.

COMPARE the technical specifications and it looks on paper like a runaway win for the Alfa GTV6, almost a no-contest affair. Beneath Guigiaro's penetrating shovel-nosed bodywork — mildly modified (and better aerodynamically) to distinguish it from that of the evergreen 'fours' — is running gear straight from the classic-car textbook: an all-alloy V6 injected engine with hemispherical combustion chambers fed by unusual overhead cam valvegear, with a single camshaft for each bank operating the inlets directly and the exhausts by cross-over pushrods; a rear-mounted five speed transaxle to give a balanced weight distribution (almost 50/50); de Dion rear suspension which reduces unsprung weight and keeps the back wheels upright regardless of vehicle weight or cornering loads; and all-disc brakes, ventilated at the front, inboard behind, again to minimise unsprung weight. The front suspension is by double wishbones and longitudinal torsion bars, the unassisted steering by rack and pinion. Given a blank sheet of paper, it would be hard to conceive a more appetising, uncompromising design than the Alfa's for a four-seater everyday express costing less than £10,000. That of the Ghia-based Ford Capri 2.8i hardly matches up to it.

The engine — Cologne's 2.8 replacing Dagenham's 3 litre which is being phased out — has less sophisticated ohv valvegear, but with its greater capacity and similar Bosch fuel injection, delivers the same 160bhp as the Alfa, and marginally more torque. Drive is through a four-speed engine-mated gearbox to a live back axle, sprung and located by two single-leaf springs *à la* cart, which are stiffer and lower than those of the 3-litre S. MacPherson struts, controlled by gas-filled dampers (as at the back), suspend the front wheels and the rack and pinion steering is one-up on the Alfa with power assistance. Braking is by a disc/drum setup, in keeping with the car's simple, almost agricultural design. It is results that count, though, not data on a spec sheet, and it's here that the tables, if not totally turned, are at least upset as the Ford exceeds expectations and the Alfa doesn't really live up to them.

You get a hint of the unexpected when first settling behind the wheel: in the Capri you're immediately at ease and relaxed in an embracing semi-bucket Recaro seat which gives a perfect driving position behind a grippy three-spoke wheel and well-aligned pedals placed at just the right height. The comprehensive, no-nonsense instruments and minor switchgear are fine too. In contrast, the Alfa is an ergonomic nightmare: despite height adjustment for the seat and steering column (telescopic movement would be far more useful here), you sit arms stretched, knees bent, Latin-ape style, behind a slippery non-grip steering wheel and a muddled facia with the speedometer isolated from the main cluster of instruments, centrally placed where they're hard to read. (Regrettably, right-hand drive cars have not been graced with the new and much better facia fitted to left hookers.) Advantage Ford.

Both engines fire immediately, idling and pulling from cold without stutter, as good injected engines should. There's little between the two on performance, either: both cars will nudge 130mph, which is cracking on a bit, and rocket to 60mph from rest in a little over 8 seconds. Both excel at the top end of the rev range rather than at the bottom (unlike the old Essex 3 litre which is the other way about), though both will lug smoothly and quite strongly from low revs, the Ford (21.2mph/1000rpm in top) predictably with rather more vigour than the Alfa (22.0mph/1000), as it's lower geared, stronger on torque and 200lb lighter. In voice and delivery, though, even Cologne's flexible, sweet-running V6 is outranked by the turbine-smooth Alfa, which in full cry emits an even more satisfying wail, not unlike that of a distant Grand Prix engine. Only its lumpy idling mars otherwise outstanding refinement. It's far from quiet but anyone who objects to a noise of such quality doesn't want an Alfa in the first place.

There's not much in it for economy either. Considering their formidable performance, both cars are remarkably frugal, the Alfa more so than the Ford in my hands (official steady-speed figures suggest that it should be), with an overall consumption of 24mpg. The Ford returned 22mpg. Driven gently, the Capri is capable of returning up to 28mpg, the GTV6 over 30, though you can subtract 10mpg from these figures for really hard driving.

With five gears to play with, the Alfa ought to have a decisive transmission advantage over the 2.8i — but it doesn't. For a start, the gearchange is a clumsy affair, with a vague gate and crunchy, obstructive engagement, no doubt because the synchromesh of the rear-mounted gearbox has the additional inertia of propshaft and flywheel to arrest. First and second are also too low — you run out of revs at the 6400rpm limit at 31 and 55mph respectively. What's more, clutch take-up is vague and concentrated at the top of the pedal's travel. The Capri, with one gear less, has a better drivetrain in several respects. It is less snatchy at low revs, particularly when ambling in town, and the change is crisp and precise. Clutch engagement is marginally better too, though as on many Fords it's irritatingly sharp and requires a delicate left foot to avoid jerky starts. With 41mph on tap in first at the 6100rpm rev-limited cut-out, and 66mph in second, the ratio spread is better than the Alfa's, giving more fluid acceleration through the gears, as well as better vigour when slogging in third or top.

The Alfa's flaws don't end there. It's steering is heavy, ponderous even when parking or rounding sharp corners where understeer can set in prematurely, limiting hairpin speeds with front-end plough. It's on fast, smooth, sweeping roads that the Alfa is at its outstanding best, with terrific grip and stability and fine placement control. Feedback through the informative steering is also excellent; you know just where you are with the Alfa. It has no untoward vices and gently forewarns of pending breakaway, if you're brave enough to reach it on dry roads. In the wet, it's easy enough to dislodge the tail under low-gear power, though such provocative oversteer is easy enough to control.

So it is with the Ford, which despite its primitive but very effective rear suspension, handles superbly and if anything clings on even more tenaciously than the Alfa, as well it might with fatter tyres (205/60 Goodyear NCTs on special Wolfrace alloy wheels, against the Alfa's 195/60 Pirelli P6 rubber on Campagnolos). Its power steering is also sharper and more precise than the Alfa's, yet still nicely weighted and by no means lacking in feel. I doubt that the Alfa would have bettered the Ford's handling and roadholding on the fast, lightly trafficked sweeping roads on which I drove the Capri back home from the Nurburgring recently, though it might well have equalled it. Where the Alfa *does* score over the Capri is on braking and ride. Despite excessive servo assistance, which calls for undue delicasy to 'feather' the pedal on and off smoothly, the brakes of the Alfa are immensely reassuring: hard road driving betrayed neither fade nor any tendency to lock the wheels. The Ford's brakes *feel* better under normal use, as the pedal is firmer and response more progressive. But a sudden panic stab can easily lock the front wheels at quite modest g, due to a pressure-limiting valve that's strongly front biased to comply with European regulations, say Ford. If this be the case, the law is an ass and the rules need changing.

Although the GTV6 can become decidely turbulent on uneven corners, rocking rudely under diagonal pitch, it normally rides more smoothly than the rather harsh and jittery Capri. In other respects, there's little to choose between the two for comfort, driving position apart where the Ford is vastly superior. Both cars have comfy, supportive seats and, by coupe standards, acceptable rear seat accommodation, though leg and headroom is cramped for anyone of above average height, particularly in the Alfa. Because it's a true hatchback, with individually folding rear seats, the Capri is also a more versatile and accommodating car than the Alfa, which has a tailgate opening on to a conventional boot. Neither car is particularly quiet, least of all at speed when wind noise begins to intrude, especially in the Capri. But what music from those V6 engines! When it comes to keeping warm and fresh, it's the Ford that excels, with its efficient heating system, tilt-and-slide sunroof (a standard fitting) and effective eyeball ventilation that does a better job than the Alfa's rather crude flap valves.

Although manufactured in Germany, the Capri 2.8i was developed in Britain by Ford's Special Vehicle Engineering operation at Dunton, Essex, under the direction of Rod Mansfield. Rod and I are old Escort Mexico sparring partners so I know better than most that he can drive a bit (he gets his off-duty kicks now from an ocean-going catamaran I believe) as well as sort cars. And there lies the key to the 2.8i's undoubted appeal: it's a car for serious drivers, well sorted by people who *are* serious drivers. Unlike the Alfa, which is sensationally good in some respects but disappointing in others, the Ford, if not flawless, is uncommonly good for a car of such humdrum origins. What's more, it's £1500 cheaper than the Alfa so I don't need to tell you where I'd put my money.

Specification	**Alfa Romeo**	**Ford**
Cylinders	V6	V6
Capacity	2492cc	2792cc
Bore/stroke	88/68.3mm	93/68.5mm
Valves	sohc per bank	ohv
Compression	9.0:1	9.2:1
Induction	Bosch injec.	Bosch injec.
Max. power	160bhp/5600rpm	160bhp/5700rpm
Top mph/1000	22	21.2
Chassis	monocoque steel	monocoque steel
Front sus.	wishbones, torsion bars, anti-roll bar	MacPherson struts, anti-roll bar
Rear sus.	de Dion, Watts linkage, coils	live axle, single leaf springs
Steering	rack and pinion	ass. rack and pinion
Brakes	disc/disc	disc/drum
Tyres	Pirelli P6 195/60	Goodyear NCT 205/60
Performance		
Max speed	130mph	129mph
0-60mph	8.5sec	8.0sec
60-80mph, top	10.3sec	8.7sec
mpg range	22-30	19-28

Alfa, left, has poorer instruments and driving position than the Ford.

When pressed hard on winding roads, this Alfa Romeo's adhesion was not as high as expected.

Six appeal

Alfa Romeo's GTV sounds an even better bet now that it has the appeal of six cylinders. JOHN BOLSTER has been trying this 130mph sports car.

The Alfa Romeo GTV has for a long time been a very attractive car in 4-cylinder form. The body is of sporting appearance, yet it provides comfortable seating for four people, while the advanced design appeals greatly to the owner who appreciates engineering finesse. Above all, the car is not too large, for excessive width spoils some of the latest sports cars.

Now, the GTV is available with a fuel-injection version of the 2.5-litre Alfa V6 engine. Quite a lot of strengthening, especially of the transmission, has taken place, but the new car retains the looks and compact dimensions of its smaller-engined sisters. I drove the machine at its introduction in Italy and liked it enormously, so I looked forward with much interest to a full road test in England. In particular, I wanted to see whether or not it suited our peculiar road conditions as well as it did those of its native land.

The technical details are fairly well known so I shall be brief. The engine is a 60deg V6 with a single overhead camshaft per bank, which operates the inlet valves directly and the exhaust valves by cross pushrods and rockers. This short power unit is also light, by virtue of its aluminium alloy block and heads, while the 4-bearing crankshaft saves friction, compared with a 7-bearing straight-six. Bosch L-Jetronic injection and electronic ignition complete a thoroughly modern design.

Technical excellence

To obtain a high polar moment of inertia, the engine is forward mounted while the combined 5-speed gearbox and final drive is at the rear. An important detail is that the clutch is on the front of the gearbox and not on the engine flywheel. With an engine-mounted clutch, the inertia of the propeller shaft would place an extra load on the synchronising arrangements during gearchanges. In the GTV6, only the two driven plates of the small-diameter clutch have to be synchronised, which is just one example of the technical excellence of the design.

The front suspension is by wishbones and torsion bars, with an anti-roll bar. At the rear, there is a trailing dead axle by courtesy of the Marquis de Dion, located laterally by James Watt's ingenious linkage and sprung on coil springs, plus an anti roll bar. The rack and pinion steering is unassisted, the low-profile tyres are Pirelli P6s, and the disc brakes are ventilated in front, inboard behind.

Wide doors make entry easy, particularly for the rear passengers, as the front seats move forward when folded. All the seats are comfortable but the driving position did not suit me. I am 6ft tall and both the steering wheel and gearbox were too far away, to engage first speed requiring an almighty stretch. My wife, on the other hand, who is only 5ft tall, sat farther forward and was quite at ease.

Greatly improved from earlier models, the gear-change has no lost motion and feels very direct in action, that peculiar 'clonk' having been eliminated. However, the springs for centralising the lever in the gate are far too strong, so that it is quite an effort to engage first and, unless one keeps up the pressure, the first to second change becomes first to fourth. This is a pity for the operation is precise and may be as rapid as one wishes,

Above: Familiar hatchback style tailgate gives easy access to the moderately-sized boot. Below: During normal driving, the handling is excellent, the effectiveness of the rear end ensuring that understeer is the chief tendency.

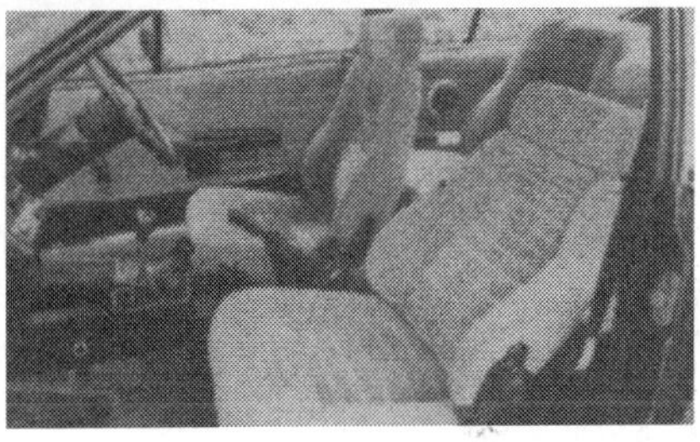

The GTV6 2.5 is aerodynamically very efficient. Inset: Interior features badly-placed instruments (left and centre) and 'Italian' long-armed driving position (right).

while the clutch is outstandingly sweet in action, though reverse was sometimes difficult to engage.

While on the subject of the gearchange, I wonder why the ratios have been chosen with first and second forming really two alternative bottom gears — I admit that I often started in second. Fifth is not a high cruising gear but it is just right for maximum speed, the car exceeding 130mph easily, though the revs are held below the ignition cutout speed. Yet, let us forget such criticisms, for cruising speeds of 110 or even 120mph are easy and effortless while the 100mph fourth may be used for miles on end when the road winds.

The engine is incredibly flexible, pulling away smoothly from 20mph in fifth gear. It revs very willingly and is quiet during normal driving, but when pressed hard it sings that glorious song that only a really good 'six' can produce. It is never noisy, but the sound that it makes is exactly right for a very refined sports car; the transmission is silent on all gears. Wind and road noise are also well subdued.

ALFA ROMEO GTV6 2.5
£9495

Specifications

Cylinders, capacity	60deg V6, 2492cc
Bore × stroke	88 × 68.3mm
Valve gear	Single belt-driven camshaft per bank
Compression ratio	9.0:1
Fuel system	Bosch L-Jetronic injection
Power/rpm	160bhp DIN at 5600rpm
Torque/rpm	157lbf ft DIN at 4000rpm
Gear ratios	0.78, 1.026, 1.345, 1.956 and 3.50:1
Final drive	4.1:1
Steering	Rack and Pinion
Brakes	Servo-assisted discs, ventilated in front, inboard rear
Tyres	Pirelli P6, 195/60 HR15
Suspension (F)	Ind. wishbones, torsion bars, anti-roll bar
(R)	De Dion axle on coil springs, trailing arms and Watt's linkage, anti-roll bar

Dimensions

Wheelbase	95ins
Track (F/R)	53.5, 53.8ins
Length	165ins
Width	65ins
Weight	23cwt 1qr
Boot capacity	7.9cu ft

Performance

Max in 5th	131mph
Max in 4th	100mph
Max in 3rd	77mph
Max in 2nd	52mph
Max in 1st	29mph
0-30mph	2.7s
0-50mph	5.8s
0-60mph	8.1s
0-80mph	13.2s
0-100mph	22.2s
50-70mph in 4th/5th	5.8/8.5s

Fuel

Urban, 56, 75mph	19.1, 36.7, 28.5mpg
Testing	22-26mpg

Above: Neat alloy wheels are one distinctive feature. Below: Six-cylinder, fuel-injection 160 bhp engine.

The steering is heavy at parking speeds but otherwise quite light, in spite of the wide tyres and absence of power assistance. The roadholding is excellent and the car is not deflected by bumps, the handling being first class during normal driving. However, when pressed really hard on winding roads it understeered fairly strongly and the cornering power was not quite as high as I had expected. The behaviour of the rear end was impeccable under all conditions.

As for the ride, exactly the right compromise has been chosen, with suspension that feels firm at low speeds but gives real comfort when cruising fast, almost irrespective of the surface. The brakes are truly excellent and while they give tremendous retardation from high speeds, with never a sign of fading, they also afford an instant stop when a drunk steps off the pavement — a virtue which all-disc systems sometimes lack.

Delightful

I was unable to test the heater properly in the hot weather but I did find that its fan was somewhat noisy. Very properly, the eyeball ventilators have entirely separate ducts but the airflow is feeble at low speeds. The instrument dials are legible but the rev-counter is mounted too far to the left — it would be better in front of the driver, beside the speedometer.

The GTV6 2.5 is a delightful car and it is difficult to put into words the pleasure that I obtained from driving it. It is exceptionally fast, even by the standards of today, but it is in smoothness and refinement that it excels. Its few shortcomings would certainly not prevent me from buying one and, who knows, perhaps the manufacturers will one day realise that Englishmen and Italians are of entirely different shapes!

Let us conclude with the most important aspect of any car nowadays. This Alfa Romeo is remarkably economical and even quite a fast driver should average around 25mpg. This is not achieved by having a high fifth gear for economy but by combining a really efficient engine with a good aerodynamic shape. There are other 130mph cars, but most of them would use a lot more petrol if they tried to keep up with this very lively sports car that bears a famous name. ■

Sweet Six

by Barry Lake

IF THERE'S ONE aspect of Alfa Romeo that is truly outstanding, it is the design of its engines. From the little flat-four Alfasud through the venerable twin-cam fours right up to the 60-degree 2.5 litre V6 designed for the Alfa Romeo 6 touring car, they are superb units; smooth, powerful and virtually indestructible.

The V6, with a single belt-driven overhead camshaft per bank of cylinders, first appeared in the "6" during 1980. In that car it has no less than *six* downdraught carburettors.

While the Alfa Romeo 6 hasn't — and possibly won't — come to Australia, the engine now has. It has arrived as the powerplant for the Alfetta GTV6, a car that will make the heart of any Alfa enthusiast pound with excitement and anticipation.

The GTV6 has the same Giugiaro-designed bodywork that graces the normal GTV, with the same black bumpers and front air dam plus side protection mouldings, but boasting as well a "power bulge" down the centre of the bonnet line. It's unmistakably the big brother of the two from a front or side view.

The V6 engine, for this car, has been fitted with Bosch L-Jetronic fuel injection and produces slighlty more power at the top end, at the cost of a small amount of torque in the mid range compared to the carburettored V6.

It's a remarkably smooth engine that has a typically Italian mechanical noise about it at idle and low rpm, but which works itself up to a healthy, pleasant-sounding exhaust burble at higher speeds. It shares with its very different relatives the ideal Alfa mix of good low down torque with an ability to rev willingly and safely right out beyond the recommended rev limit if necessary. It's one of those rare engines that is just as happy idling along as it is singing its heart out at high speed.

The GTV6 is considerably faster than the four-cylinder version, even though our test car was so new it failed to approach the best times shown in overseas magazines.

Our GTV6 was 1.1 seconds faster than the four over the standing 400 metres and 0.7 seconds, 1.1, 1.8, 2.1, 2.9 and 3.6 seconds faster from rest to 60, 80 100, 110, 120 and 130 km/h respectively.

And it did that with far less effort. The gearchange, never good on the GTV, was tight as well as vague on this very new car. It's a shame the engineers didn't put some effort into improving this on the new car. It has the same basic gearbox, mounted at the rear in unit with the differential, but with different ratios to match it to the characteristics of this engine.

The clutch has dual driven plates to handle the extra power of the larger engine, giving it a strange feel in that the pedal must be fully depressed to release both plates completely, yet almost fully raised to re-engage them. Changes on the move, however, can be made without fully depressing the pedal.

The brakes have been uprated. The front discs are slightly larger in diameter and ventilated. They work superbly, as we've come to expect from Alfas. The Pirelli P6 tyres mounted on smart-looking alloy wheels contribute to the braking ability as well as to the handling.

The steering and handling are very similar to the GTV. Although the GTV6 is 104 kg heavier than the GTV, it doesn't seem all that much heavier in the steering while the handling is still basically strong understeer. The wishbone/torsion bar front suspension and de Dion rear axle on coil springs with Watts linkage location are both pretty much the mixture as before.

Many of the criticisms of the GTV remain, therefore. I still don't like the driving position — steering wheel and gear lever too far away, pedals too close for comfort — nor the lack of headroom.

Likewise, the instruments are still badly placed, with tacho and other important gauges over in the middle of the dash panel.

Nor has the gearing been altered on the painfully slow window winders.

The front seats are comfortable. The rears are adequate for children up to teenage and will hold two adults in relative comfort for shorter trips. Luggage space is good for a car of this type.

Fuel economy seems like a strong point. We had the car for too brief a period (we hope to get another one soon) to do extensive consumption checks but with a figure of 13.5 litres/100 kms including the performance runs, it would appear that figures of 12.0 L/100 or better would be quite within the car's capabilities.

The high point of driving this car is feeling the willingness of the engine. It is quite remarkable in its versatility and provides better performance with fewer gear changes than its four-cylinder brother. At a price. At the time of writing, we don't have a figure on this, but the car will surely be on the wrong side of $30,000.

But that's a price quite a few people will be prepared to pay in order to slip behind the wheel of one of the best-looking cars on Australia's roads today. It is also clearly the fastest Alfa Romeo currently available and this alone will give it appeal for the Alfa enthusiasts. Many of them will buy the GTV6 and will love it with such a blind passion we will be called fools for daring to criticise it on any count. □

ALFA ROMEO Alfetta GTV6
ROAD TEST DATA
(As tested)

ENGINE

Location	Front, mounted fore and aft
Cylinders	60 degree V6
Bore X Stroke	88.0 X 68.3 mm
Capacity	2492 cm^3
Fuel injection	Bosch L-Jetronic
Compression Ratio	9.0 to 1
Fuel Pump	Electric
Valve Gear	Belt-driven Single OHC per bank
Claimed Power	117.6 kW at 5600 rpm
Claimed Torque	219.9 Nm at 4000 rpm
Maximum Recommended Engine Speed	6000 rpm
Specific Power Output	47.2 kW/litre

TRANSMISSION

Type	Five-speed manual, rear-mounted
Driving Wheels	Rear
Clutch	Two dry plates

Gearbox Ratios

Gear	Ratio	km/h-1000 rpm	Max Speed (at rpm)
First	3.500	7.9	47 (6000)
Second	1.956	14.2	85 (6000)
Third	1.345	20.6	124 (6000)
Fourth	1.026	26.9	161 (6000)
Fifth	0.780	35.4	198 (5600)
Final-Drive ratio			4.1 to 1

SUSPENSION

Front	Independent by wishbones and, torsion bars with anti-roll bar
Rear	De Dion axle located by trailing arms and transverse Watt's linkage with coil springs
Wheels	Alloy 6.0J X 15
Tyres	Pirelli P6 195/60 HR15

BRAKES

Front	265 mm discs
Rear	250 mm discs

STEERING

Type	Rack and pinion
Turns, Lock to Lock	3.5
Ratio	n/a
Turning Circle	10.0 metres

DIMENSIONS AND WEIGHT

Wheelbase	2415 mm
Front Track	1373 mm
Rear Track	1352 mm
Overall Length	4190 mm
Overall Width	1650 mm
Overall Height	1330 mm
Ground Clearance	110 mm
Kerb Weight	1184 kg
Weight to Power	10.1 kg/kW

CAPACITIES AND EQUIPMENT

Fuel Tank	75.0 litres
Cooling System	12.0 litres
Engine oil capacity	n/a
Battery	12V 66 AH
Alternator	65 Amps

CHECKLIST

Alloy Wheels	Yes
Adjustable Steering	Yes
Air-conditioning	Yes
Carpets	Yes
Central door locking	No
Clock	No
Intermittent Wipers	Yes
Laminated Screen	Yes
Petrol-filler lock	Yes
Power Steering	No
Power Windows	No
Radio	No
Tape Player	No
Rear-window Wiper	No
Remote outside mirror adjustment (electric)	Yes
Sun Roof	No
Tachometer	Yes

FUEL CONSUMPTION

Average for Test	13.5 litres/100 km

ACCELERATION

0-40 km/h	2.51 seconds
0-60 km/h	4.26 seconds
0-80 km/h	6.46 seconds
0-100 km/h	10.12 seconds
0-110 km/h	11.81 seconds
0-120 km/h	13.58 seconds
0-130 km/h	16.23 seconds
Standing 400 metres (130.8 km/h)	16.41 seconds

PRICE

As tested	(approx.) $32,000

Forest fire

Engineering advances bring Japanese sports cars ever closer to European standards. Can Colt and Toyota now compete with the pace set by Alfa Romeo's classic GTV6 coupe?

One of the lessons which car manufacturers are having to learn is that unless the scales of production and sales are of multi-national proportions, there is precious little profit to be made in the business of making bread and butter cars.

The result has been that those car makers not in a position to attack the major markets have pushed their products up market, and have raised prices correspondingly to collect the rich pickings which are available in cars which sell at around the £10,000 mark. The importers of the three cars in this test are in just this position. Toyota and Colt are heavily restricted in the number of cars they are allowed to sell in Britain, and Alfa Romeo whilst not under any quota control have not either the right product or manufacturing capacity.

The logical move for all three was, therefore, to import a small number of cars with Porsche-style exclusivity and of course Porsche-style price tags.

When Alfa Romeo began to sell the GTV6 2.5 in Britain last year the price soon leaped up to over £9000, a significant increase on the amount previously charged for the lesser two-litre models. Toyota say that they will be importing only a few of their similar Supras – about one per dealer, and Colt, as a small importer will ensure the new Sapporo Turbo to be a rare sight on the roads. The £8000 price of the Capri Injection which has a similar specification can give some idea of the mark up on these £10,000 coupés.

THE CARS

Our three cars comprise a group of traditional front engined rear-driven coupes, although within that configuration the Alfa Romeo offers a slight variation in the interests of improved weight distribution and therefore in theory improved handling. The complete gearbox and clutch unit is situated in unit with the rear differential.

Under the bonnet lies a very over-square, all alloy V6 engine of 2492cc which produces 160 bhp, an increase of 30 bhp over the two-litre GTV. Unlike the twin-cam four cylinder Alfa engines the V6 engine uses just one cam per bank of three cylinders, and unlike the similar Alfa 6-saloon engine which uses a multiple carburettor set-up, the GTV has Bosch L-Jetronic fuel injection.

Front suspension is by double wishbone and torsion bars, and a sophisticated deDion axle is used at the rear. Brakes are servo-assisted discs all round, inboard at the rear. The steering is unassisted rack and pinion, and alloy wheels are shod with Pirelli P6 tyres.

The GTV's eccentric facia which for so long was criticised for having its instruments ridiculously placed – the main instrument pod being in the centre of the facia – has now been changed, but it is worth pointing out that the revision is carried out in Britain by the importers and not at the factory in Italy. Basically the dials are regrouped and the facia retrimmed with a digital clock included in the centre of the new layout.

There are two versions of the GTV using this handsome coupé bodyshell. The two-litre model is

priced at £8150 and the V6 car costs a mighty £10,250, which makes it the most expensive car of the trio.

Toyota's Supra is an expensive version of the cheaper Celica but the relationship tends to be somewhat nominal as although the bodyshell is similar, it is longer. It sits on a wheelbase four inches longer, and is differently styled at the front and rear, including pop-up headlamps.

A straight-six cylinder engine powers the Supra, displacing 2759cc and with twin belt driven overhead camshafts. It produces 168 bhp and is a development of the single cam unit found in the Crown saloon. Fuelling is by electronic injection and the transmission is by a five-speed gearbox. No auto option is offered.

Suspension at the front is by MacPherson struts, and the rear of the car is independently sprung by coil springs and trailing arms. Anti-roll bars are fitted to front and rear. Steering is by power-assisted rack and pinion and braking is by discs all round aided by a servo. Alloy wheels are a standard fitting.

It goes without saying that equipment levels are high on the Toyota but an interesting fitting is that of air conditioning. There is just one version of the Supra and it costs £9888.

The Sapporo Turbo uses the same four-cylinder 1997cc engine as powers the ultra-quick Lancer Turbo, producing 170 bhp – the same output as the Lancer – and which features the Colt's 'Silent Shafts' which are intended to cut down on vibration. Turbocharging increases the power output by a massive 59 bhp over the normally aspirated engine. Electronic fuel injection is fitted to complement the turbo power. Transmission is by a five-speed gear box.

The suspension set-up is very similar to that of the Toyota with MacPherson strut suspension at the front and independent trailing arms at the rear. Steering is power-assisted recirculating ball. Like the other two cars, disc braking is used all round with a servo. Tyres are 70 series low-profile rubber shod on alloy wheels.

A relative scarcity of supply will allow the Colt Sapporo Turbo to sell for £10,158 which is an increase of no less than £2274 over the cheaper model. It is worth bearing in mind that the striking looking Starion 2000 Turbo sells for £11,734, not a tremendous amount more than the Sapporo.

PERFORMANCE

A slight power disadvantage coupled with a few pounds less in weight sets the Alfa closely against the other two cars which are near identical in their power-to-weight ratio, so prior to testing it was not clear which car would prove to be the fastest – no doubt an important factor to those considering buying any of the three.

Nevertheless it is the Alfa which beats the Colt and Toyota on paper and also feels significantly quicker. The V6 engine really is incredibly smooth and responsive, and in comparison with the carburettored Alfa 6 the fuel-injected version is no less than a transformation: 60 mph is reached in 7.6 secs and 100 mph is made in 23 seconds – fast by any standards. It makes just the right noises too, a beautifully deep V6 note roars out right the way through the rev range.

It must be said though, that the very rapid acceleration is achieved partly due to rather ridiculously low gearing. First gear for example is almost unnecessary as the Alfa will move off with virtually no transmission judder in second. At the other end of the rev range the V6 engine is revving its head off and though giving a lot of high speed acceleration response, it seems strange that Alfa Romeo could not rely a little on the engine's good torque output and raise the final drive ratio.

The other gripe on the Alfa is the gearchange. The quality and ease of the linkage on these rear-mounted gearboxes is infamously variable (from poor to appalling, say the cynics) but even on the best examples, of which we had one, it is awkward and requires a real concentrated effort to achieve a smooth change. Why don't Alfa Romeo take a Porsche 924 to pieces to see how it should be done properly?

The Toyota's straight-six gives luxuriously effortless performance, and answers the criticism that the two-litre versions of the car were basically underpowered. It has immense torque; the engine will pull academically low speeds in fifth. However it is not as willing to rev as hard as might be expected from a twin cam unit and actually gets a little rough over about 4000 rpm, and particularly so at the 6500 rpm rev limit. It's a fast and generally refined engine but lacks the excitement of the Alfa's V6.

With the gearbox mounted in the conventional position aft of the engine there is not the same problem in getting a good change as there is with the Alfa, and indeed the movement is slick with a close gate and precise selection of the gears. Overall gearing is much more suited to cruising than it is on the Alfa – second gear reaching 65 mph is an ideal ratio for serious overtaking, strings of slow moving caravans on winding roads, for example.

Initially the Colt actually feels quite slow compared with the other two cars, seemingly lacking much punch, but on paper the figures are easily comparable with 60 mph reached in 8.1 secs and 100 mph in 24.5 secs. The turbocharger certainly hasn't got the massive kick that one experiences with the blown Lancer, feeling altogether a much milder boost. It comes into effect at about the same engine speed – about 3000 rpm but gives more of a strong surge forward than a punching kick. However in second gear the turbo cuts in at about 30 mph and usefully continues until around 50 mph, and the effective turbo power band spreads from 40 to 70 mph in third – ideal second and third gear overtaking speeds. Like the Toyota, the Colt is excellent for darting in and out of gaps in the traffic. The engine itself is quiet and mechnically refined but it cannot match the six-cylinder silkiness of the others in the test.

The gearchange is probably the best of the group – light with a precise gate and very well placed in relation to the driving position. Maximum speeds in each gear are in fact similar to the Alfa but there is not the same feeling of being under-geared.

All three cars have high top speeds, the Celica nudging ahead to make 130 mph, just one mph ahead of the Alfa and a clear 10 mph more than the Sapporo. It goes without saying that all will easily cruise at well in excess of the legal limit.

HANDLING AND RIDE

Does the Alfa Romeo's sophisticated rear suspension and ideal weight distribution result in ideal handling? Up to a point, yes. Cornering is well balanced – though with perhaps just a touch too much understeer – and the rear end clings on well with power on or off. There is a flaw though, and that is that the chassis does not feel especially taut under cornering due to excessive and very unsporting amounts of body roll and a feeling of slight corkscrewing through bends.

The steering feels heavy at slow speed but lightens up as speeds increase. There is plenty of response in the system – helped by the absence of power-assistance – but some drivers may find it picks up too many small road irregularities and feeds them back. Parking is a real chore due to the steering effort required, not helped by the wide Pirellis.

Where the Alfa's suspension does pay off is in the ride quality. It smooths out road bumps with a well damped but firm ride, without any of the harshness one might find in, say, a Ford Capri Injection. The all-round disc braking system pulls the car up very effectively though needing a firm push. Once used to the Alfa, drivers will probably prefer this type of sensitivity. As the car draws up to a halt there is other a slight rocking sensation felt through the suspension.

The Celica has well engineered suspension and it is good that the extra weight of the Supra's engine and extra equipment doesn't seem to have done it any harm. There is not the excessive understeer one might have expected from such a long-nosed coupé, and in fact the Supra turns into bends very well, the feeling of quickness being enhanced by steering which has only three turns from lock to lock. The tail end is a different story though, with massive slides on tap, brought about either by lifting off the power, or booting heavily down on the throttle. The steering is quick but is too light to allow any feel at all when cornering. It's a luxury in town though – parking is a delight.

Damping quality is a curious mixture of firmness and spongy springing. On corners there is no tautness and the Supra feels woolly, but the ride is firm at the same time, and despite the independent rear suspension the coils feed back some crashing over bumps. Braking is very powerful and reassuring but perhaps just a shade over servo assisted.

The Colt possibly offers the best set-up of the three cars. It corners well without too much roll and the rear end clings on well.

Steering unfortunately, is hopelessly light and provides absolutely no feel at all. A little less power assistance would be well in order. Ride quality is the firmest of the three but is still good being comfortably damped. Only a slight joggliness – a common Japanese trait – lets it down. Front and rear discs provide secure stopping.

ACCOMMODATION

None of these three can truly be described as a four-seater car – but the extent of rear seat space in these two-plus-twos ranges from the good to the very poor.

The Alfa's seat comfort is poor for a £10,000 car. The front seats are quite small and the low roofline of the coupé body means that headroom is at a premium even for smaller drivers, and those over six feet might well have a real job fitting in comfortably. A seat height adjuster helps but the driving position is generally unsatisfactory – the steering wheel, which is adjustable for angle but not for reach, is too far away in relation to the pedals.

Rear seating is fairly cramped in the Alfa with headroom even more scarce than it is in the front. The seats themselves are well shaped and will just about accommodate average sized occupants but without great comfort.

The Colt, due to its more saloon-like body proportions, offers the best accommodation. Front seats are comfortable and supportive, the driving position is good and there is enough head room. It's also clearly the best for rear seating too, as the average adult can easily fit in and not feel cramped. Indeed there is as much room in the rear of the Sapporo as one would find in many proper four-door saloons.

In contrast, the Supra is two-plus-two coupé with only only nominal seating in the rear. Adults must sit with their necks bent and legs splayed in order to fit into the rear. Like the Alfa the seats are actually quite well contoured, but few people will be able to take advantage of this.

Front seating is good with masses of adjustment forward and rear, although headroom could be a close shave for taller front occupants. A very unusual feature of the front driver's seat is that the pressure of lumbar support can be increased by pumping up a little 'bladder' to the left of the driver's seat, and the air in the seat released by pressing three buttons on the side of the seat back. It sounds ridiculous but in fact it works well and a driver can easily use it without any distraction whilst on the move.

The Supra offers the best luggage space because it has a tailgate and individually folding down rear seats giving a useful if rather modest platform due to the low roofline and room consuming rear differential. In contrast the Alfa is little more than useless in this respect. The car has a small tailgate which when opened reveals only a meagre size of boot, and irritatingly the rear seats cannot be folded down as one might expect when lifting a hatchback.

With its saloon/coupé styling mix the Colt has conventional boot and as such its ultimate accommodation is not as flexible as the Toyota's, but what boot space there is, is reasonable, offering an unobstructed if rather shallow space. The rear seat backs fold down individually which is useful if long objects need to be carried.

ALFA ROMEO GTV6

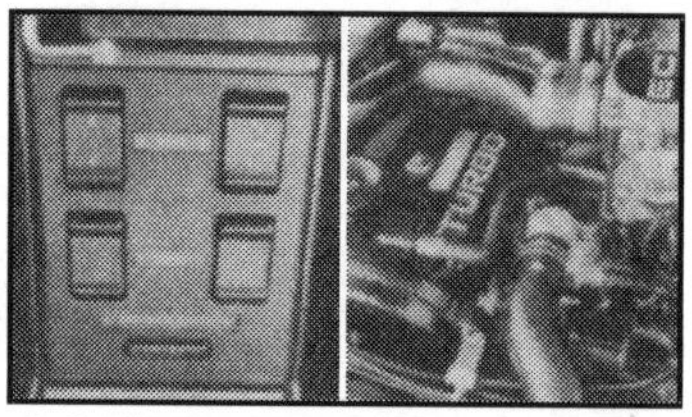

COLT SAPPORO TURBO

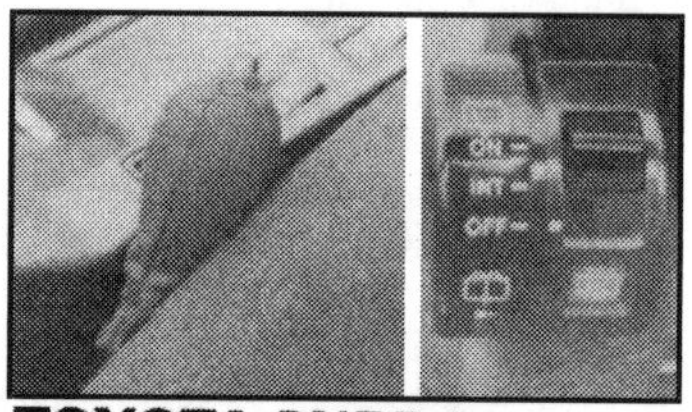

TOYOTA SUPRA

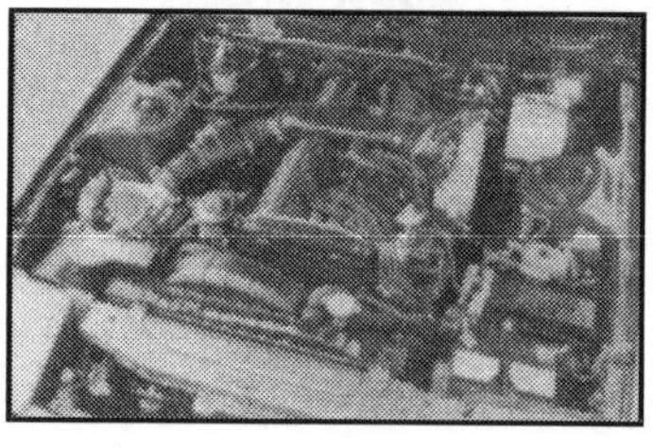

LIVING WITH THE CARS

Testing the Toyota Supra during a particularly hot and humid summer spell allowed us to test the standard air conditioning to the full. It gives a fine cool atmosphere when fully turned up and is circulated by a quiet fan, which really does make life pleasant on a hot and sticky day.

This completely contrasts with the Alfa's ventilation which is virtually non-existent. There is a fan fitted but it seems to do little more than make a loud irritating whirring noise even on the slower speed. The only way to get proper ventilation is to open either the windows or the sunroof. The Colt's system falls between the two cars in efficiency. It has the usual Japanese regulating controls and

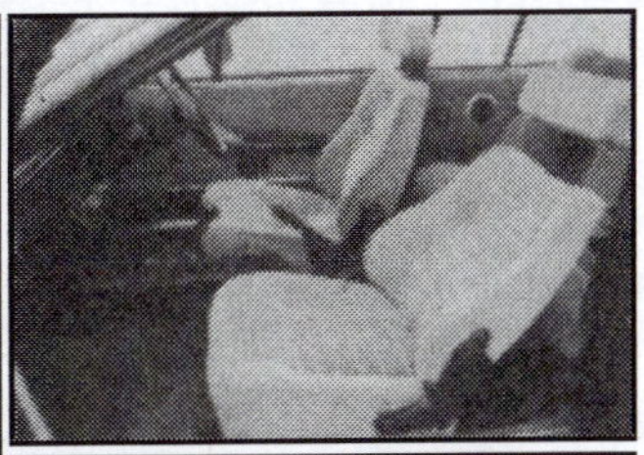

Alfa's handling is still too stodgy for a sports coupé, with roll and understeer. Engine is silky delight and facia is improved but interior remains cramped

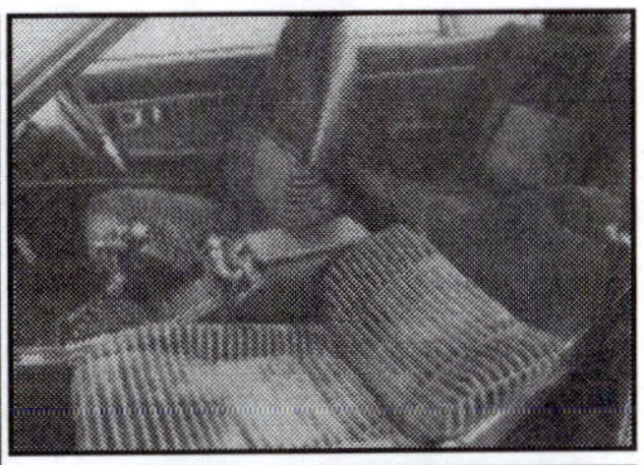

Colt corners well but over light power steering spoils handling. Turbo engine gives smooth performance, interior is roomiest of all but the plain facia is poor

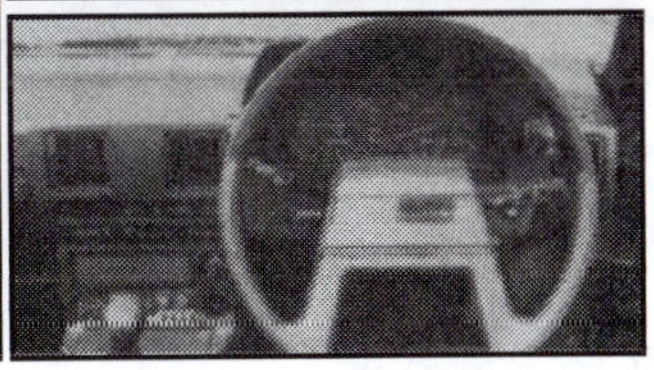

Light steering also mars Supra handling but car is agile to drive. Twin-cam engine is fast but not classic, seats are clever but the facia is depressingly cheap

gives a good blast of air and is fine by ordinary non air-conditioned vehicle standards.

Expensive coupés such as these cars should have an invitingly designed facia which is also clear and easy to read. As already mentioned, the Alfa Romeo has its disastrous facia converted in England – European left-hand-drive models already have an improved dashboard – and while this is passable it is certainly not one of the car's strong points. The speedometer is small and it is a bit too obvious that the gaps where the central dials were have been covered over (albeit neatly) and filled with an Alfa Romeo badge and a digital clock. The steering wheel is unsatisfactory, being too large and finished in slippery polished wood. But it does have a useful height adjuster fitted.

The Supra has a big clear facia which is easy to read and looks reasonable but the finish to the switchgear looks horrible and some of the switches – particularly the rotary ones – look like the cheapest-after-market accessories. The steering wheel in particular is of an unappealing design and is also far too big bearing in mind that there is power assistance in any case. The Sapporo's facia is functional but has the plain, mass produced Japanese look, not matching the price tag of the car. The turbo boost gauge in particular looks feeble, tending to flick on and off, which of course gives no indication of boost pressure building up in the turbo system.

The Alfa is the main culprit for high noise levels due to the howling six, whereas the main source of intrusion on the other two cars is wind noise, with the Toyota's wing mirrors being a source of turbulence. All three cars have attractively decked out interiors but we thought the light tan trim of our Alfa would show up the dirt particularly badly after a while.

COSTS

We weren't expecting especially good fuel consumption figures from these three big coupés but relatively gentle driving brought a best of 28 mpg from both the Alfa Romeo and Toyota. The Alfa's worst figure worked out at 23.0 mpg which is creditable for such a big engine.

The Colt's worst was 20.2 mpg and its best figure was 23.5 mpg; the heaviest consumption went to the Toyota which sank to just over 19 mpg. All the cars' fuel consumption returns came as quite a pleasant surprise as we were expecting little more than an average of 20 mpg from any of them.

Parts prices are not going to be cheap on these cars and particularly expensive is the exhaust system on the Alfa which costs £276. However the quoted labour times are fairly short for all the cars.

Servicing intervals for all three cars are all long, the Supra and Sapporo needing attention at 10,000 miles, and the Alfa Romeo needing a major service at 12,000 miles. All three require a small service at the halfway mark. There are 184 Alfa dealers, 240 Colt agents, and 205 for the Toyota.

Make sure you will be able to insure any of these three cars properly. Both the Colt and Toyota will fall into group seven at least, and the Alfa is likely to be even more expensive, in group eight, according to many insurers, which will mean very big bills.

It's perhaps worth considering which other cars are available at around this price. With a very similar specification is the Capri Injection at £8125, the Porsche 924

Group test: Alfa Romeo GTV6, Colt Sapporo Turbo, Toyota Supra

ALFA ROMEO GTV6

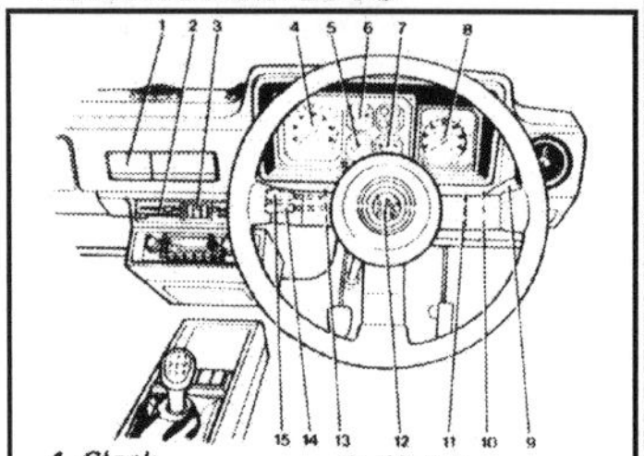

1. Clock
2. Heater
3. Fan
4. Rev counter
5. Fuel gauge
6. Oil pressure
7. Temperature
8. Speedometer
9. Wipers, washers
10. Rear fog lights
11. Choke
12. Horn
13. Hazard warning
14. Indicators
15. Lights

COLT SAPPORO TURBO

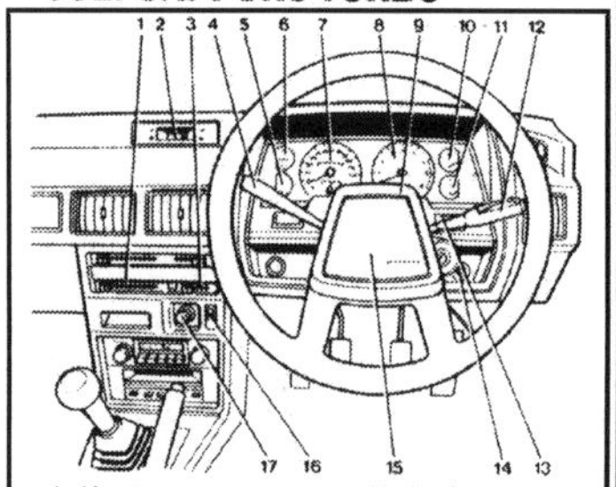

1. Heater
2. Clock
3. Fan
4. Wipers, washers
5. Fuel gauge
6. Battery condition
7. Speedometer
8. Rev counter
9. Turbo boost
10. Oil pressure
11. Temperature
12. Lights/indicators
13. Rear fog lights
14. Panel rheostat
15. Horn
16. Heated rear screen
17. Lighter

TOYOTA SUPRA

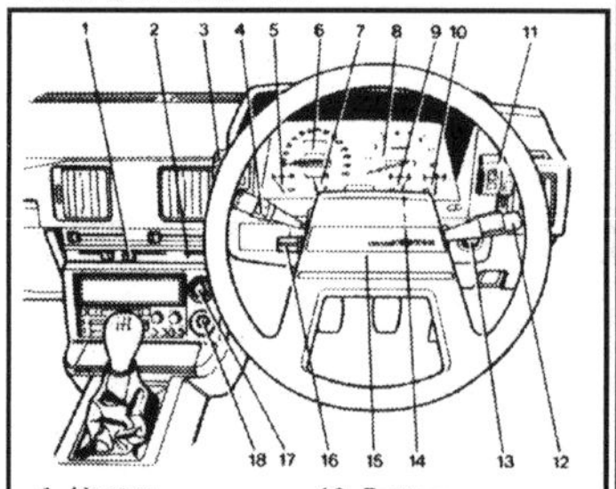

1. Heater
2. Fan
3. Rear wipe, wash/heated rear screen
4. Wipers, washers
5. Oil pressure
6. Speedometer
7. Temperature
8. Rev counter
9. Fuel gauge
10. Battery condition
11. Cruise control
12. Lights/indicators
13. Panel rheostat
14. Hazard warning
15. Horn
16. Column adjuster
17. Lighter
18. Air conditioning

CAR	Alfa Romeo GTV6 2.5	Colt Sapporo 2000 Turbo	Toyota Celica Supra 2.8i
PRICE	£10,250	£10,158	£9998
Other models	1	1	2
Price span	£8150-10,250	£7884-10,158	£6604-9888
PERFORMANCE			
Max Speed (mph)	129	120	130
Max in 4th (mph)	105	105	126
Max in 3rd (mph)	83	89	96
Max in 2nd (mph)	56	57	65
Max in 1st (mph)	33	37	37
0-30 (sec)	2.3	2.9	3.0
0-40 (sec)	3.9	4.6	5.1
0-50 (sec)	5.3	6.1	6.8
0-60 (sec)	7.6	8.1	8.8
0-70 (sec)	10.1	10.9	10.9
0-80 (sec)	13.4	13.7	14.3
0-90 (sec)	16.6	17.5	18.0
0-100 (sec)	23.0	24.5	23.0
0-400 metres (sec)	16.0	16.2	16.7
Terminal speed (mph)	87	89	88
30-50 in 3rd/4th/5th (sec)	3.9/5.2/7.0	5.8/9.1/12.2	5.6/6.8/9.2
40-60 in 3rd/4th/5th (sec)	4.1/5.6/8.5	4.8/7.5/10.3	5.3/6.8/8.8
50-70 in 3rd/4th/5th (sec)	4.9/7.1/9.4	4.4/6.7/9.2	6.0/7.4/9.5
60-80 in 3rd/4th/5th (sec)	6.0/7.4/11.0	4.8/7.0/9.9	6.3/7.7/10.0
SPECIFICATIONS			
Cylinders/capacity (cc)	V6/2492	4/1997	6/2759
Bore stroke (mm)	88.0 x 68.3	85.0 x 88.0	83.0 x 85.0
Valve gear	ohc	ohc	ohc
Compression ratio	9.0:1	7.6:1	9.2:1
Induction	injection	injection/turbo	injection
Power/rpm (bhp)	160/5600	170/5500	168/5600
Torque/rpm (lbs/ft)	157/4000	181/3500	169/4600
Steering	rack/pin	PA/rec ball	PA/rack/pin
Turns lock to lock	3.5	3.7	3.0
Turning circle (ft)	33.0	32.8	32.8
Brakes	S/Di/Dr	S/Di/Di	S/Di/Di
Suspension front	I/Wi/Tor	I/McP	I/McP
rear	de D	I/S/TA/C	I/STA/C
COSTS			
Test mpg	23.0-28.6	20.2-23.5	19.3-28.3
Govt mpg City/56/75	19.1/36.7/28.7	22.4/37.1/29.1	21.5/34.1/27.4
Tank galls (grade)	16.5(4)	13.2(4)	14.3(4)
Major service miles (hours)	12,000(34)	12,000(4.1)	10,000(2.8)
Part costs (fitting hours)			
Front wing	£138.00(12.5)	£97.52(0.8)	£55.12(3.5)
Front bumper	£82.00(0.4)	£53.95(0.5)	£211.85(4.1)
Headlamp unit	£34.50(0.1)	£74.30(0.4)	£14.12(0.6)
Rear light lens	£38.00(0.3)	£34.09(0.6)	£37.69(0.6)
Front brake pads	£6.80(0.5)	£47.35(1.0)	£11.93(1.7)
Shock absorber	£24.80(0.4)	£34.73(1.3)	£65.89(2.1)
Windscreen	£94.00(1.8)	£60.20(4.0)	£60.74(3.0)
Exhaust system	£276.00(0.8)	£117.55(1.4)	£142.36(1.5)
Clutch unit	£301.00(2.3)	£71.30(3.1)	£63.06(3.5)
Alternator	£73.70(0.9)	£40.00(0.5)	£123.43(0.7)
Insurance	8	7	7
Warranty	12/UL	12/UL	12/UL
Anti-rust	5 yr	none	none
UK Dealers	184	200	205
EQUIPMENT			
Central locking system	No	No	Yes
Electric windows	Yes	Yes	Yes
Laminated screen	Yes	Yes	Yes
Tinted glass	Yes	Yes	Yes
Petrol cap lock	Yes	Yes	Yes
Power steering	No	Yes	Yes
Adjustable steering column	Yes	Yes	Yes
Radio/Cassette	Yes	Yes	Yes
Seat height adjustment	Yes	No	Yes
Rear seat belts	No	No	Yes
Sunroof	Yes	No	Yes
Rear wash-wipe	No	No	Yes
Headlamp wash-wipe	Yes	Yes	No
DIMENSIONS			
Front headroom (ins)	33	35	35
Front legroom (ins)	37-42	34-41	33-40
Steering-wheel-seat (ins)	17-22	10-19	11-19
Rear headroom (ins)	31	33	32
Rear kneeroom (ins)	22-27	25-33	30-38
Length (ins)	167.7	178.2	181.9
Wheelbase (ins)	94.5	99.6	102.9
Height (ins)	52.5	53.0	51.8
Boot Load height (ins)	35	35	32
Overall width (ins)	65.5	66.0	66.3
Int. width (ins)	53	58	53
Weight (cwt)	23.8	24.6	24.9
Towing weight (cwt)	19.6	28.4	19.7
Boot capacity (cu. ft)	10.6	16.8	10.9/21

KEY. Valve gear: *ohc, overhead camshaft.* **Steering**: *rack/pin, rack and pinion; rec ball, recirculating ball; PA, power assistance.* **Brakes**: *Di, discs; Dr, drums; S, servo assistance.* **Suspension**: *I, independent; C, coil springs; Tor, torsion bar springs; Wi, wishbones; McP, MacPherson struts; deD, de Dion axle; STA, semi-trailing arm location.*

Lux costs £9582, and the Datsun 280ZX and its 2+2 variant cost £9354 and £10,317 respectively.

VERDICT

These three coupés differ slightly in their appeal and therefore it is not easy to pick an obvious winner. For sheer fun and excitement the Alfa Romeo is the obvious winner and many drivers will be instantly charmed by its sporting heritage, the deliciously responsive V6 engine, and its stunningly attractive good looks.

But what buyers in search of a sophisticated coupé will find extremely irritating on the Alfa is the gearchange – which as we said can vary in quality from car to car – the lack of front headroom and the cramped rear, the indifferent facia and awful ventilation, and the high noise levels. Ultimately one must decide whether the numerous faults are compensated by the wonderful performance. We are tempted to say they are, but £10,250 is a lot of money.

The Toyota came as quite a pleasant surprise as this company seems to be undergoing a regression in their quality of styling and engineering with other models at the moment. But the Supra is quick, quiet and refined, and the changes to the body styling improve its looks over the rather ordinary Celica.

Probably its main drawback is that rear seat accommodation is minimal and as such it can only really be considered to be a two-seater coupé. Exclusivity tends to be a contradiction in terms with Japanese cars but it is quite likely that the Supra will remain a relatively rare sight on British roads.

The Colt is probably the most 'sensible' car of these three. It easily has the best accommodation with proper rear seats; performance and handling are good if not outstanding, and it has all the modern motoring conveniences found in Japanese cars. Character is not its strong point though. Styling is just a little bland and the interior – especially the facia – is very ordinary and forgettable, and not really the part in a car costing this much. Altogether it's a good car but against the other two it does not exude such a strong personality or appeal.

Provided you have a clear idea of the type of motoring you wish the choice is simple: the Alfa is the true sports car, the Supra is the long distance cruiser, and the Colt is the good all rounder. And a few minor improvements would make the Alfa best on all three counts.

Group test: Alfa Romeo GTV6, Colt Sapporo Turbo, Toyota Supra

WHAT'S NEW? for '83

Italian Love Affair?

YOU CAN'T BE OBJECTIVE about cars like the GTV6. Either you love them or you hate them. All Alfas are drivers' cars and with that they demand and they give — a lot. All the Italian virtues of positive, accurate steering, fuss-free handling, a lovely gearbox and sexy Italian looks holding their age very well. In fact the comparison to Italy's Sophia Loren makes for a pleasant few moments lost daydreaming. Like the actress, the car has been around for some time but seems to be improving with age. Unfortunately like the actress the car is getting more expensive everyday and the challengers to the throne get stronger.

It used to be that the Veloce only got delivered with the straight four DOHC dynamo. The engine was, and still is, one of the best two-litre engines around but the driver had to be careful chosing combat venues other than the Boulevarde grand prix.

Where once the handling was the saving grace for the car, now a new heart transforms the Veloce into a force to be reckoned with. Where once even the most dyed-in-the-wool Alfa fanatics found themselves wishing for more grunt, more *'forza,'* now and their prayers are answered. In spates!

The engine in the GTV6 is the most powerful unit built by Alfa Romeo. It is a 60° all-alloy V-6 displacing 2492cc. The compact layout generates less torsional vibrations than an inline six and makes it possible to utilise a very short crankshaft supported on four main bearings. Alfa capitalised on its racing experience to determine the bore and stroke ratio of 88mm by 68mm. The oversized valves (for the capacity) have been designed in a tight 46° V angle to ensure maximum volumetric efficiency. The local unit develops maximum power at 5600 rpm — down slightly on the European version. Torque is also marginally lower than the overseas version at 212 Nm (versus 220) at 4000 rpm.

To determine the block size Alfa engineers took pointers from the Formula One flat-12 engine. The entire timing system is built around two overhead camshafts, the actual dynamics being conceived from computer analysis. At 140 kmh only half power is used so the motor is left with a solid reserve of torque. At 47.4 Kw per litre (64.2 bhp) the GTV6 unit is right up with latest efficiency figures.

To control the GTV6 power plant Alfa selected a L-Jet-tronic fuel injection system. The unit uses a computerised flowmeter to determine the optimum fuel-air mixture based on data including engine speed and cylinder head temperature. Injection fueling provides accurate metering of the mixture at all times so that the ratio of air to petrol is always optimized under all driving conditions.

Alfa has finally introduced the slingshot single overhead cam six shaped in a 60° block that, given half an opportunity will melt down the new GTV's fat low-profile radials in nothing flat. By combining the new 117 Kw engine with the Sprints' good handling and four wheel discs, Alfa has played right into the hands of every frustrated non-Porsche-driving sporty car *afficionado.* Stand up those drivers not presently doing time for Saturday-night escapades or Sunday-afternoon violations. Where is that neighbourhood 911 that always seems to appear halfway round a corner?

In addition to the V6, mechanical changes of note include 15 x 6 Campagnolo magnesium wheels and chunky 195 section P6s. The wheels make way for slightly larger vented front discs. With all this extra power and performance, springs and shockers are upgraded along with a new twin-plate clutch that gets the ponies back to the 5-speed transaxle.

Anyone at home with the four will feel right at home with the V6. All the reminders are there. The usual Italian standard-ape driving position — great if you have a long torso and short legs — which limits headroom for us taller drivers. The seats are hip and shoulder hugging with good lumbar support that makes a difference on those interstate jaunts, following the Alfasud Series around Oz. Even though the car is a new model there is instant *deja vu.* One wonders if this can be off-putting to a non-Alfanatic potential customer. Forking out near — on twenty-seven big ones might make them think about something that is more recognisable as a new car.

Sitting in the car, the sights and sounds are pure Alfa. The characteristic exhaust note is still there. Perhaps a little more subdued and smoother from the extra pots exercising. Trying for first gear the driver is instantly reminded to pander to the slightly recalcitrant gearbox. The trick here is to move the shift lever toward second before engaging it in first. We didn't get an opportunity at the launch to fully test the macho V6, but all indications point to some impressive driving experiences to come . . .

Flexibility is a long suit of the GTV6. The torquey engine will let the lazy driver (is there such a thing with an Alfa?) potter around town in fifth gear, barely above idle, and never protest. Acceleration figures of high fifteen second quarters and seven second 0-100 kmh times have been recorded by other drivers. Top speed should be into the 210-220 bracket.

A few counterpoints arose at the launch. First, is the car such great value at $27,000? If you compare it to the 944 Porsche, which Alfa does, then it holds up very well in all areas. But if you look at it like a first time buyer might — it is not a cheap car. Unsophisticated buyers who might just want a posey, flashy sports car would probably opt for the considerably cheaper Starion Turbo, or Datsun 280 ZX. Others wanting a good sports car at a realistic price would jump into the sturdy, faithful RX-7. Even your BMW 323i or all-new Prelude are quick cars and good value — the new Honda excellent value at $16,500! In all fairness though, none have the breeding, prestige or balls-out performance of the GTV6.

Another point, which many gloss over these days, (company executives included) is that even in the Eighties the single most important factor in new car sales is still Design-/Styling.

It's here that the Alfa, as good as it is, falls down a little. The shape is a proven Guigara design that is a modern day classic, yet it is dated against truly aerodynamic shapes now appearing. Its Cd figure is only 0.39 in bespoilered form. To 'untraditional' Alfa buyers, seeing the car for the first time, they just might think it's another Italian car living in the past. If they don't drive it, they will have missed something special. But these days *conquest* sales are so important that a car has to be able to attract un-traditional customers. This, then is the dilemma . . .

COMPARATIVE TEST

ALFA GTV 6 AND BMW 535i

We don't see them as alternatives so much as divergent ideals. But if they were, what a choice to spend time on....

FAST cars are fun: and that's the main motivation behind this off-beat comparative test, assessing the Alfa-Romeo GTV6 *alongside* the BMW 535i. It's off-beat because its format breaks the guidelines set for these reports when we started comparative testing in TECHNICAR in 1975.

We decided then that the cars had to be priced fairly closely and of similar type, competing for the same slice of the market; and viewed in this way, our Alfa-BMW project would be a non-starter, because how could you pit a R23 500 2.5 litre sports coupe like the Veloce "six" against a 3.5-litre four-door sports sedan costing R4 000 more?

You couldn't. But on the other hand they are the two fastest cars built in South Africa and despite the big difference in engine capacity, they've been locked in battle on local race tracks throughout the past year.

It would obviously be fun to test them alongside each other, and we felt that the fact that they represented such contrasting concepts of a high performance car might work to our advantage: it should throw each of them into clear perspective.

So that's out aim and this particular test has to be viewed in this light. It's not Alfa versus BMW – which would be grossly unfair; its two different approaches to fast motoring, developed for different types of enthusiast. Just this once, the prices are incidental.

First, let's examine their backgrounds, then look at the engineering, which represents some of the best that Germany and Italy have to offer, embodying the national characteristics of Italian flair and German efficiency.

Both cars represent a flag-ship, high performance development of an established model range. The sleek, road hugging GTVs have established a phenomenal reputation as ultra rapid sports cars with

competition rostrums, when driven by people of the calibre of Arnold Chatz, who have exploited their cornering power to offset any shortfalls in straight line speed.

Extremely fast overall

As a two-litre, even with the twin-cam Alfa motor, the GTV was extremely fast *overall* — from point to point — but could fall a little short in a sprint down the straight. But when Alfa introduced their compact, cammy 2,5-litre V6 motor early last year, they slotted it neatly into the Veloce body, giving it extra muscle to go with its agility.

With strengthened drivetrain, uprated suspension and almost unrivalled balance (which owes something to a rear-mounted transaxle) the GTV6 became one of the fastest cars available and our April 1982 road test established a top speed of over 200 km/h — within spitting distance of Alfa's claim (based presumably on ideal conditions) of 205.

At that time, the 3,5-litre BMW 535i had just set up a speed of 206, on test with CAR, and the two cars had no challengers for the joint title of the fastest locally built models on the road.

The fact that the top speeds were so close, despite the litre gap in engine capacity, was easily explained by the BMW's larger, heavier sedan body and overdrive gearing, which gave it a figure of 46,4 km/h per 1 000 r/min in top. But at that level of performance, the difference was of academic interest to most drivers.

Refined punch

However, the 535i, which had superseded the tearaway 530 by offering the refined punch of the 3,5-litre "six" used in the 735i in the compact but roomy 5-Series bodyshell with luxury trim and equipment, then underwent a significant change. Instead of the five-speed overdrive gearbox used overseas, South African buyers could opt for a close-ratio sports type transmission with a direct fifth gear. As the 3,07 final drive was retained, the overall gearing dropped by a dramatic 19 per cent to 37,3 km/h per 1 000 r/min in top — to within 2 km/h of the Alfa's figure of 35,6.

The Alfa's dash treatment is classic but unco-ordinated, with speedo binnacled separately from rev-counter and other instruments. The wheel is wood-rimmed and alloy spoked, to match the pear-shaped gearshift knob, and pedals are beautifully spaced and proportioned. Seats are well shaped (left, below) and fold for rear seat access. The fuel injected 2,5-litre V6 (right) produces 116 kW at 5 600 r/min, with abundant torque.

This made the car even more of a bombshell on the race track, with searing acceleration times and a top speed (established in a retest for our January 1983 performance issue) of 220,7 km/h. But, of course, there was a price to pay in terms of fuel consumption, which increased (for example) from 8,2 to 10,3 litres/100 km at a steady 100 km/h.

In this form, the 535i also became an even hotter contender in the high performance car market, though it was never intended for volume production and is rarely seen in any BMW showroom: there is a constant waiting list of impatient buyers.

Both cars embody the intangible but breathable mystique of their illustrious forebears — an aura that stems from their classic grille treatments and flows right through the machines. They both express the highest credos of their makers, for they are built to carry their owners far and fast in safety and with limitless pleasure — provided, of course, that they can handle such potent cars competently. . .

To a certain point, the specs run reasonably parallel: six-cylinder, fuel-injected cammy motors driving the rear wheels through five-speed transmissions, all-disc braking, light alloy wheels shod with low profile tyres, adjustable steering columns (the Alfa's for *angle*, the BM's for *length)*, and electrically operated windows and side mirrors — one on the GTV, two on the 535i. However, as we'll see, they're different in almost every other way.

Exceptional torque

The Alfa's 2 492 cm^3 60° V6 develops 118 kW at 5 600 r/min and its output, while generous, has been deliberately restrained in favour of an exceptional spread of smooth, hefty torque, peaking at 213 N.m at 4 000 r/min — a characteristic found in previous classic units of this make, along with an eagerness to rev.

It drives through a rear-mounted alloy cased transaxle to obtain optimum mass distribution, and therefore improved handling and traction, and the suspension is also typical of the marque: expensive but efficient, with torsion bars and stabiliser in front and a semi-independent De Dion axle at the rear. Used in conjunction with a Watts linkage and anti-sway bars, it keeps the rear wheels parallel in all situations and lays the foundation for phenomenal handling powers.

BMW's 3 453 cm^3 straight six is, of course, the race-bred, highly developed unit used in the 735i, which develops 160 kW at 5 200 r/min, with an electronically governed limit of 6 200 r/min and 310 N.m of torque at 4 000 r/min.

For this motor, this is a moderate stage of tune and the power delivery is silkily smooth but extraordinarily muscular: a combination that makes for exceptional driver enjoyment, in a car of this type. Despite the mass of 1 470 kg — 260 kg more than the Alfa's — its power-to-mass ratio is extremely high (108,8 W/kg) and in performance, this is the nearest thing to an ultra refined, four-wheeled surperbike we have ever driven.

The fully independent suspension is essentially an uprated development of the standard 5 Series arrangement, with coil struts and stabiliser bar in front and semi-trailing arms and a second stabiliser at the rear. Bilstein gas shocks and spring rates are set to achieve an unusual combination of firm high speed hand-

The BMW 535i fascia is among the best planned of modern cars — co-ordinated and functional — but that small, thick rimmed racing wheel (from the M1) spoils the driver's view of some instruments. The Recaro seats (below, left) are superb and the race-bred fuel-injected 3,5-litre straight-six (right) is as used in the 735i, developing 160 kW at 5 200 r/min with 310 N.m of torque available at 4 000 r/min.

ling and executive style comfort.

Any enthusiast knows that you can't compare sports cars on a basis of performance figures alone. It's the pleasure and satisfaction they yield to the driver that counts most and while one man gets his kicks from cannon-ball acceleration, another would prefer optimum balance, response and faster cornering — which might make a "slower" car just as quick and twice as fun, for him, over a twisty route.

Significantly better tune

But, for the record, the 535i with sports gearbox scorches from 0 to 100 in just 7,9 seconds (see tables) — and this figure, and the 220 km/h top speed, are so far ahead of the figures recorded on our previous test a year earlier that they cannot be accounted for by the gearing change alone. The latest car must have been in significantly better tune...

With a power to mass ratio of 97,2 W/kg, and a litre less muscle under the bonnet, the GTV certainly can't match figures like these; but its 0-100 km/h time is only 10,8 seconds and it burns a lot less fuel achieving it: at a steady 100 km/h, its consumption is only 7,83 litres/100 km — 24 per cent below the figure for the BMW.

Our 100 km fuel run, in which the two cars run closely coupled over a varied route, sharing the same conditions and recording the same times, confirmed the substantial gap in consumption, with the Alfa recording 8,8 litres/100 km and the BMW, 13,8.

Let's move on to their characters. In truncated definitions, the Alfa is an excitable, super responsive sports car, while the BMW is a compact, hyper-powered express, fitted with luxury trimmings — even central locking — and with room for a family. While they are both easy to drive, the BMW requires much more expertise when you're really pressing, because its tremendous power can break out the rear end fairly easily, particularly on a wet road, and if it's overdone, the car then requires catching...

If you had to select a really fast car for a novice, the Alfa would be a sound choice, because it can be hurled round corners with unschooled abandon, provided you keep your nerve and maintain control. The BMW, on the other hand, would be risky: it requires a judicious foot on the throttle, which becomes a vital balancing element in the cornering process, when motoring fast.

Blood bond with driver

Perhaps the dominant virtue, in the GTV, is the hardest to describe. It seems to establish a blood bond with the driver, so that as you weave and snort through the corners, you feel motor, transmission and tyres working in feed-back pulses. That gives a deep, gut sense of satisfaction that livens up your life...

It seems to *flow* across the tarmac like some low slung mechanical leopard, imparting an incredible sense of balance and clinging power, but still so classically rorty, you expect to smell hot engine oil. You don't, but if you did, you'd accept it along with the raking glass overhead (which can be a little trying on a hot day), the shortage of head room and the cramped rear seats: it's an insignificant price to pay for what this car has to offer.

And among the pleasures is the noise: a deep burble on start-up, rising to an eager snarl with each blip on the throttle and settling to a throaty rasp at cruising revs.

Rakish, low and superbly styled, the GTV has uncompromising two-plus-two seating and an interior that is a cross between an Italian fighter aircraft and the classic racers that Fangio used to drive, brought partially up to date. The instrument layout is comprehensive but individual, with a 240 km/h speedo visible through the wood rimmed wheel (which can become slippery), a big rev-counter binnacled separately, along with sundry gauges and warning lights, over to the left and that curved green fan switch governing a ventilation/heating system which is not the most efficient around but does a reasonable job.

Floppy gearchange

The wooden, pear shaped gear knob falls comfortably to hand, the pedals are beautifully spaced/shaped but the gearchange requires knowing and responds to skill. At first, it's loose, almost floppy as you explore the linkage to that remotely mounted box; but once you develop the light, scooping motions and are used to the twin-plate clutch which has its action at the top, you obtain precise, quick cog swops with a sense of satisfaction.

If the Alfa suggests a leopard, the BMW is a charging bull — but a bull with unobtrusive refinement and impeccable manners. Driven with aplomb and the required degree of skill, it responds with predictable discipline; with restraint, it's so docile the wife would never guess at the fury lurking underneath the bonnet.

Although 5 Series styling and wolfish grille hint at sporting capabilities, underlined by the low stance and fat tyres, the only overt clue to the blistering performance comes from a modest but potent model tag at each end of the car, prefaced by M (for Motorsport) and the company's racing colours.

It has the Recaro seats, sophisticated air-conditioning, programmable mo-fi and electronic service indicator and active check control found on other top flight BMWs. But one item that seems oddly out of place in the restyled, cleverly co-ordinated fascia is the fuel consumption indicator: surely the last thing a 535i driver would want to watch, even if it didn't dampen his exuberance!

The ride is just a trace firmer than other 5 Series cars and pays off at really high speeds, when the car remains rock steady and can be steered with relaxed precision. The M1 racing wheel is nice to handle but obscures vital gauges and we'd prefer the standard steering wheel (available as an option) for its practicality.

The motor is smoother and quieter than the Alfa's and the superlative performance is delivered with Teutonic efficiency, rather than passion. The mechanical refinement tends to isolate the driver, making both motor and suspension feel relatively remote — an impression that is heightened by the power assisted steering.

The competition clutch is firm and deep and the short, stubby gear lever has a short but deliberate action that requires a firm throw for peak performance. The gate is unusual, with reverse forward and to the left, almost opposite first, making 2/3 and 4/5 available in lightning-straight movements.

So if price is little object and you want a truly fast car, one of these should prove your ultimate. If you've a tribe to lug around or you cover a fair distance on dirt, the BMW has practical advantages, even if its character and equipment have no decisive appeal. As we said at the beginning, we don't see them as alternatives so much as divergent ideals. But if they were, what a choice to spend time on! ●

SPECIFICATIONS

Alfa GTV 6 — R23 500
BMW 535i — R27 500

	ALFA ROMEO	BMW
Make	ALFA ROMEO	BMW
Model	GTV6	535i
PERFORMANCE FACTORS:		
Power/mass (W/kg) net	97,2	108,8
Frontal area (m^2)	2,21	2,41
Km/h per 1 000 r/min (top)	35,6	37,3
INTERIOR NOISE LEVELS:		
Idling	63	54
60 km/h	74	70
80 km/h	76	72
100 km/h	79	75

(Measured in decibels, "A" weighting, averaging runs both ways on a level road with car closed.)

ACCELERATION FROM REST (seconds):	ALFA ROMEO	BMW
0-60 (km/h)	5,0	3,7
0-80 (km/h)	7,4	5,4
0-100 (km/h)	10,8	7,9
1 km sprint	30,9	28,5

OVERTAKING ACCELERATION:

	3rd	4th	Top	3rd	4th	Top
40-60 (km/h)	3,4	4,8	6,8	2,6	4,2	5,3
60-80 (km/h)	3,4	4,8	6,6	2,4	3,8	5,4
80-100 (km/h)	3,4	5,0	6,8	2,3	3,9	5,4

(Measured in seconds, to true speeds, averaging runs both ways on a level road, car carrying test crew of two and standard test equipment.)

MAXIMUM SPEED (km/h):	ALFA ROMEO	BMW
True speed	201,5	220,7
Speedometer reading	205	233

Calibration:

Indicated:	60	80	100	60	80	100
True speed:	55	74	93	51	70	89

FUEL CONSUMPTION (litres/100 km):	ALFA ROMEO	BMW
60	6,45	8,68
80	7,14	9,31
100	7,83	10,31

(Stated in litres per 100 kilometres, based on fuel economy figures recorded at true speeds.)

FUEL CONSUMPTION (OVERALL):	ALFA ROMEO	BMW
Overall	8,9 litres/100 km	13,8 litres/100 km

(Recorded over two laps of a 49 km varied road circuit with the two cars travelling closely together and with one mid-point driver change.)

BRAKING TEST (From 100 km/h):	ALFA ROMEO	BMW
Best stop	3,4	3,3
Worst stop	3,8	4,0
Average	3,6	3,44

(Measured in seconds with ten stops from true speeds at 30-second intervals on a good bitumenised surface.)

GRADIENTS IN GEARS:	ALFA ROMEO	BMW
Low gear	1 in 2,5	1 in 1,9
2nd gear	1 in 3,5	1 in 2,6
3rd gear	1 in 5,4	1 in 3,6
4th gear	1 in 7,8	1 in 5,7
Top gear	1 in 10,3	1 in 8,0

(Tabulated from Tapley (x gravity) readings, car carrying test crew of two and standard test equipment.)

GEARED SPEEDS (km/h):	ALFA ROMEO	BMW
Low gear	45	53
2nd gear	80	81
3rd gear	116	110
4th gear	152	157
Top gear	200	195

(Calculated at engine power peak — 5 600 r/min for the Alfa and 5 200 r/min for the BMW.)

ENGINE:	ALFA ROMEO GTV6	BMW 535i
Cylinders	60° V6	In-line 6
Fuel supply	electronic injection	Bosch L-Jetronic
Bore	88,0 mm	93,4 mm
Stroke	68,3 mm	84,0 mm
Cubic capacity	2 492 cm^3	3 453 cm^3
Compression ratio	9,0 to 1	9,3 to 1
Valve gear	o-h-v, single o-h-c (per bank)	o-h-v, single o-h-c
Main bearings	four	seven
Fuel requirement	98-octane (Coast) 93-octane (Reef)	98-octane (Coast) 93-octane (Reef)
Cooling	water	water
ENGINE OUTPUT:		
Max power I.S.O. (kW)	118	160
Power peak (r/min)	5 600	5 200
Max usable r/min	5 700	6 200
Max torque (N.m)	213	310
Torque peak (r/min)	4 000	4 000
TRANSMISSION:		
Forward speeds	five	five
Gearshift	console	console
Low gear	3,50 to 1	3,717 to 1
2nd gear	1,956 to 1	2,403 to 1
3rd gear	1,345 to 1	1,766 to 1
4th gear	1,026 to 1	1,236 to 1
Top gear	0,780 to 1	direct
Reverse gear	3,000 to 1	4,233 to 1
Final drive	4,10 to 1 (3,198 overall)	3,07 to 1 ZF LSD
Drive wheels	rear	rear
WHEELS AND TYRES:		
Road wheels	alloy, sports	alloy, sports
Rim width	6,0 J	6,5 J
Tyres	195/60 HR 15 radials	200/60 VR 14 radials
BRAKES:		
Front	265 mm vented discs	280 mm vented discs
Rear	250 mm inboard discs	272 mm discs
Hydraulics	dual circuits, anti-lock at rear, vacuum servo	dual circuits, anti-lock at rear, vacuum servo
Handbrake position	between seats	between seats
STEERING:		
Type	rack and pinion	ZF variable ratio power-assisted
Lock to lock	3,5 turns	4,6 turns
Turning circle	10,1 metres	11,1 metres
MEASUREMENTS:		
Length overall	4 260 mm	4 620 mm
Width overall	1 664 mm	1 690 mm
Height overall	1 330 mm	1 425 mm
Wheelbase	2 400 mm	2 636 mm
Front track	1 373 mm	1 422 mm
Rear track	1 352 mm	1 460 mm
Ground clearance	122 mm	140 mm
Licensing mass	1 210 kg	1 470 kg
SUSPENSION:		
Front	independent, torsion bars, stabiliser bar	independent, coil struts, stabiliser bar, gas shocks
Rear	De Dion triangulated axle, coils, anti-sway bars, Watt's linkage	independent: coils, semi-trailing arms, stabiliser bar, gas shocks
CAPACITIES:		
Seating	2+2	4/5
Fuel tank	75 litres	70 litres
Luggage trunk (net)	340 dm^3	440 dm^3

TESTER'S COMMENTS

Body styling, design and finish	Completely contrasting. BMW has standard 5 Series four-door bodyshell and only sign of its explosive performance is the neat model tag at each end, prefaced by M (for Motorsport) and the company's racing colours. The Alfa is a low, rakish and superbly streamlined two-door sports coupé with two-plus-two seating and exotic appeal. Both cars are fitted with expensive imported alloy wheels and low profile, wide section imported radial tyres.
Minor collision protection	Alfa has deformable padded plastic bumpers and heavy side protective mouldings. BMW has fully padded steel wrap-round bumpers and protective side mouldings.
Boot design and size	Alfa's 340 dm^3 boot accessible through rear hatch, supported by centre strut beneath security cover which forms wide parcel shelf when down. BMW's 440 dm^3 boot is deep and unobstructed and can be locked by the central locking system or separately, with the key.
Seating, upholstery and access	Alfa has fully reclining sports type bucket seats in front and shaped rear seats, upholstered in high quality material and providing excellent support. Front seats fold for rear seat access. BMW has exceptional Recaro seats in front and unusually comfortable rear bench with easy access through wide doors.
Dashboard design	Alfa's fascia is unusual: unco-ordinated, at first glance — a mixture of classic tradition and modern instrumentation which will please Alfisti. The BMW has a revised, highly co-ordinated treatment incorporating active check panel (in the roof), a service interval indicator and fuel consumption indicator. Alfa's instruments include 240 km/h speedo, 7 500 r/min tachometer red-lined at 6 000, oil pressure, water temperature and fuel gauges, digital clock, and warning lights for turn indicators (one!), handbrake, charge and choke. BMW's display includes 240 km/h speedo, 7 500 rev-counter (red-lined at 6 500 but governed at 6 200!), a central warning light, a digital clock and warning lights for brake pad wear, brake fluid, handbrake, oil pressure and charge. It has two turn indicator lights.
Heating and ventilation	BMW has sophisticated electronically controlled air-conditioning/heating system with dial-a-temperature facility and multi-direction distribution controls. Alfa's heating/ventilation system is relatively simple, incorporates powerful heater but barely adequate ventilation system.
Under-bonnet accessibility	Both engine compartments crammed with machinery. These are not cars for ordinary DIY mechanics to tackle.
Control layout and character, gearchange	Both cars have quick shifts but Alfa's requires knowing and BMW has unusual gate pattern. BMW has heavy competition clutch but Alfa's twin-plate clutch is reasonably light. BMW's power assisted steering provides effortless low speed control — when Alfa's wheel becomes rather heavy. Alfa has three stalk controls, BMW two. Alfa's steering column adjustable for angle BMW's for length.
Handling and ride	Both cars are precise and safe at high speed but BMW's enormous power can break out rear end relatively easily and the car requires experienced handling. Alfa is noisier, has harder ride but provides fantastic mechanical feed-back and "feel" which is rewarding to the enthusiast.
Petrol tank/cap	Alfa has lockable petrol cap, BMW's petrol flap is governed by central locking system.
Horns	Both cars have powerful twin horns — BMW's operated by pads in steering wheel, Alfa's by central button.
Lights	Both cars have powerful twin halogen headlights. BMW also equipped with front and rear foglights.
Guarantees and service intervals	Alfa warranty covers two years or 100 000 km on engine, gearbox and differential, two years on paintwork (fade or rust) and one year on other mechanical parts. BMW warranty covers 12 months or unlimited distance, with six years rust warranty.

ENGINE SPEED

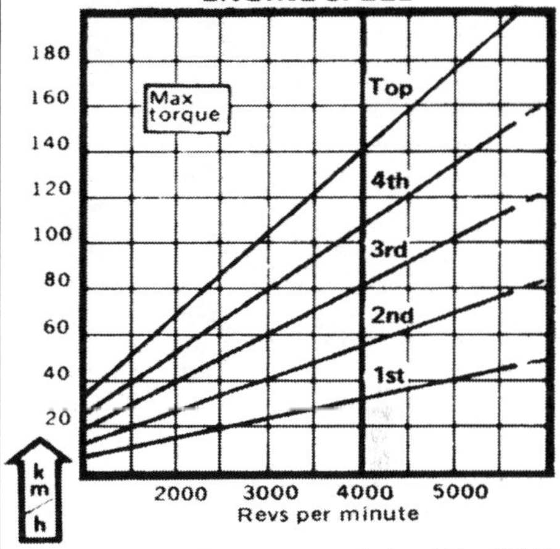

Above: engine speed graph of the Alfa GTV6 (top pic). Below: engine speed graph of the BMW 535i.

ENGINE SPEED

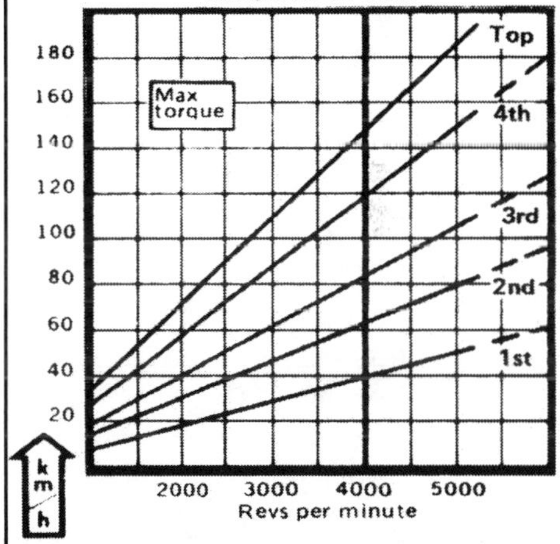

REAL PERFORMANCE —WITHOUT A TURBO

Mention 'GTV' to a sports car enthusiast and a gleam will usually appear in his eyes as his mind conjures images of a fast, powerful and sleek machine. It is such an image that has surrounded the most sporty of the Alfa Romeo range for many decades (although there are some Japanese manufacturers who have used the same alphabets for their sportier models).

After many years of being restricted to the 2000 cc sports car segment, Alfa Romeo suddenly found that they had ample justification to produce an even bigger GT model than the GTV 2.0. Market surveys showed that Alfa Romeos were going to be in greater demand in the coming years in Europe and an additional model in the upper range would be useful, especially since they already had the Alfa 6 saloon as the flagship.

The huge engine of Alfa 6 was the logical choice for powering the new car which we now know as the GTV6 2.5. The classic 2-litre engine in the GTV 2.0 had reached the limit of its enlargement and a flagship for the GT range would have to be unique. Thus was born the GTV6 2.5, featuring the very oversquare 2492 cc, all-aluminium V6 (60 degrees) powerplant of the Alfa 6.

Technically, the V6 engine has an interesting cylinder head design which uses single overhead camshafts (per bank) that operate the inlet valves directly while operating the exhaust valves with a cross-over pushrod arrangement. The entire system was arrived at only after a sophisticated design study using a computer. The choice of valve layout (with a narrow angle of 46°) is a solution which originated from the experience gained in the days of the GTAM and Type 33 racing cars. Thus, it has been possible to achieve a hemispherical combustion chamber with only one belt-driven camshaft per bank.

The vee configuration offers advantages of compactness and good balance, the latter point being particularly important to sportscar engines. The better balance — and smoothness — comes about as a result of less torsional vibration on the crankshaft than would be experienced in an in-line 6-cylinder engine. Also useful for a car like the GTV6 is the willingness to rev to high engine speeds.

Front end is sharply tapered with a deep airdam that is heavily flared for maximum effect.

The intriguing shape of the GTV is continued in the GTV6, albeit with slight changes here and there. It is a highly functional shape that looks fast even when the car is standing still.

In collaboration with Bosch, an electronic fuel-injection system was developed for the car. The system is as sophisticated as it is efficient, thanks to the use of a microprocessor which calibrates the mixture of fuel and air for each combustion chamber in accordance with a number of parameters.

The system uses a flowmeter which, by means of a special throttle valve, sends a voltage signal to the microprocessor (mounted inside the car). This signal, which varies according to the airflow, determines the exact duration and amount of fuel to be injected through the electromagnetic injectors at each port. Besides the flowmeter signal, the microprocessor also takes into consideration engine speed, coolant temperature and a series of other variables that are critical to attaining optimum performance.

Electronic fuel-injection has therefore been wise for an engine such as this, especially an Alfa Romeo engine. Because the 'brain' takes into account the condition of the engine, minute adjustments can be made to suit variations caused by wear and tear, carbon deposits, etc, thus enabling optimum performance at all times.

It is therefore not surprising to find that, even with a moderate compression ratio of 9:1, the power output of the GTV6 engine is an impressive 117.6 kW (DIN) at 5600 rpm with peak torque of 212.9 Nm at 4000 rpm. Such high outputs put the GTV6 in the upper ranks of sports cars but it is the exceptional power-to-weight ratio of 10.3 kg/kW which beats all others in this class of cars. Notable in Alfa Romeo philosophy is the fact that, at 140 km/h, only half the maximum power available is utilised, giving ample power reserve. Torque too is available in generous reserves, giving good engine flexibility.

Traditionally, Alfa Romeos have sported 5-speed gearboxes not so much for economy reasons but for performance. The GTV6 continues not only this tradition but also another, that of having the gearbox and clutch assembly mounted at the rear. The transfer of these two fairly heavy components to the rear has helped to achieve that much sought-after 50:50 balance in weight distribution which does wonders for handling and roadholding. This is particularly important too as the V6 engine is a heavy unit and taking some weight away benefits the driver as the steering won't be so heavy.

Naturally, the various components in the drivetrain have had to be beefed up to accommodate the higher torque output. Besides significantly strengthening the gearbox, propeller shaft and driveshafts, the clutch is of a twin dry plate type with more robust thrust bearings and a large master cylinder. cylinder.

The suspension layout will be familiar to enthusiasts of the marque as it is essentially the same as that used under the Alfetta and GTV 2.0. The independent front suspension comprises torsion bars with location by means of double wishbones. Telescopic dampers with progressive action control shock transmission while a thick anti-roll bar keeps the GTV6 on an even keel during cornering.

Under the rear end, adjacent to the transaxle, is a de Dion solid rear axle favoured by Alfa Romeo for various reasons, one of which is the lower unsprung weight exerted. A Watts parellelogram linkage provides the necessary lateral restraint whilst

The truncated rear end allows air to leave the car with less turbulence.

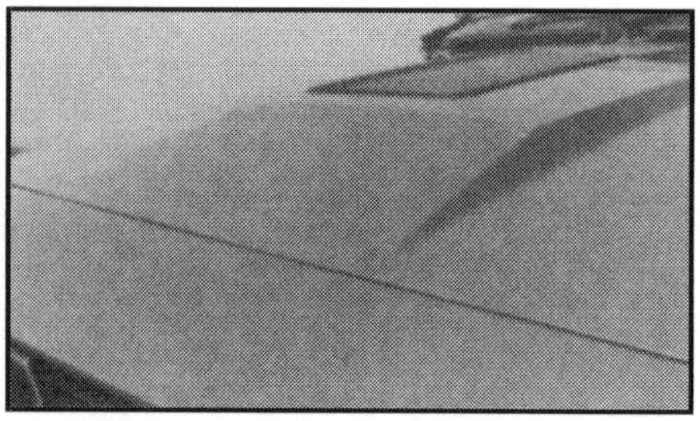

How do you tell the new GTV 2.0 from the GTV6 2.5? By this bulge on the bonnet, commonly referred to as a 'power bulge'. Black panel on top is, however, a detraction from functionality of the styling.

The sharply sloping rear screen ends with a panel that is slightly curved to act as a spoiler, exerting downforce on the rear end for better high-speed stability.

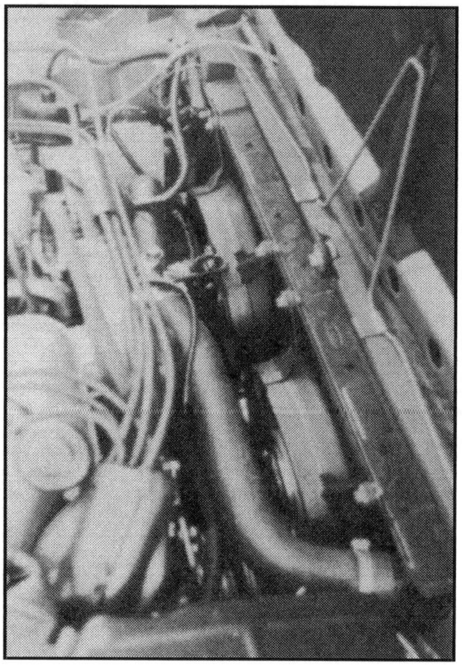

Two fans are used to ensure that the GTV6 has adequate cooling, especially with an a.c. fitted.

longitudinal location comes from trailing arms. Due to the geometry of such a suspension layout, the rear wheels are always kept perpendicular to the road and Alfa Romeo feel that such a condition offers the best in terms of handling and roadholding.

An anti-roll bar is also fitted for the rear, along with heavy-duty coil springs and telescopic dampers.

Well aware of the performance levels of their cars, Alfa Romeo have always made sure that adequate stopping power is available, hence the usual fitment of disc brakes on all four wheels.

With the kind of speeds the GTV6 can touch, the 'mandatory' 4-wheel disc brakes are given extra stopping power by the use of ventilated 267 mm diameter discs in front. Dual hydraulic circuits for safety are standard, along with a big vacuum servo unit for boosting brake pressure. A regulator in the rear brake lines ensures no wheel lock-up during the weight-transfer in hard braking.

A car like the GTV6 would be done a great injustice if a set of alloy wheels weren't fitted so Alfa Romeo decided to include Campagnolos as standard items. Sized at 6J × 15, these rather flat alloy wheels are shod with expensive 60-series tyres (usually Pirelli P6) with a size of 195/60HR15. You will note that the wheel diameter of 15 inches is larger than normal and this is simply because the use of such low-profile tyres makes it necessary to increase the wheel diameter in order to maintain the correct rolling circumference.

From mechanicals, let's look at the aesthetics of the GTV6. When the Alfetta range was introduced in the mid-'70s, the GT/GTV coupe versions took on sharp and somewhat intriguing body lines that were a stark contrast to the fondly remembered GTV of the early '70s. The new shape, as dramatic to us then as the new Sierra is today, still manages to turn heads with its 'aggressive' lines. As to be expected of such sleekness, the aerodynamic efficiency is high with a Cd of 0.39.

While the Germans are good at functional styling, they aren't always able to combine this with a certain flair that the Italians have. The GTV6 front end is a good example of aggressive looks blended with functional form and no frills.

Two pairs of circular quartz-halogen headlights flank an all-black grille with only the Alfa Romeo motif giving visual relief. The projecting polypropylene bumper with 6 km/h impact resistance is painted the same as the body colour, giving an integrated look. The integral look is continued underneath the bumper where a prominent airdam of

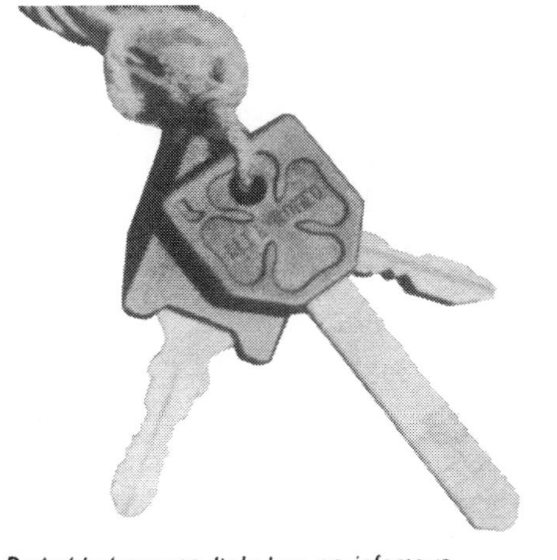

Probably because Italy has an infamous reputation of being a place where car thefts are rampant, the locks of the GTV6 require special keys.

The speedometer alone faces the driver while the tachometer and ancillary gauges are situated towards the centre in a separate binnacle. This arrangement is heavily sports-oriented but isn't practical in a road-going car

high-impact plastic is mounted. Its bottom end has a 'lip' of sorts which enhances airflow diversion around the sides to minimise lift. To supplement the volume of air entering via the grille, extra air is allowed to enter the engine bay by means of grilled vents moulded in the airdam.

Those who read our July issue may remark that the new GTV 2.0 now looks like the GTV6 so how does one go about differentiating the two? Simple — look on the bonnet. The GTV6 has a rectangular 'power' bulge topped by a black trim piece meant, presumably, to resemble an access hatch. To some, this little item might enhance the car's sporty looks but we feel that it seriously detracts from the car's functional styling — especially since it serves no purpose.

The 'no nonsense' look of the GTV6 is carried on in the side view with absolutely no signs of chrome on the bodywork. For effect, the window frames are painted black, and this theme is carried along on the drip channels, stale air vents at the rear quarters, fuel filler cap and even the door handles. A black plastic moulding also runs along the lower edge of the body, between the wheel arches, giving both aesthetic enhancement and protection from corrosion.

From a maximum height of 1330 mm, the roof line curves back down towards a truncated rear end with the rear screen and hatch slightly recessed between two ridges. To obtain some downforce on the rear end, the trailing edge of the hatch tilts up in what is now commonly referred to as a 'ducktail'.

If the looks of the GTV6 aren't enough to convince you of its nature, the very act of getting into it will. The low roofline means some crouching and the low seats put you quite close to the ground. And then, of course, those well padded wraparound seats start the adrenalin flowing.....

The driving position isn't too bad, ergonomics-wise, although there seems to be the usual bias towards people with long legs. Some effort has been made to minimise the common criticism about Italian cars being made for strangely-shaped people by having a tilt-adjustable steering column and height-adjustable seat in the GTV6. The latter adjustment, in particular, will be welcome by anyone who has sat in a GTV because the bonnet slopes so sharply that most people just cannot see the nose.

The large glass area, nice in cold climates, can be a bit bothersome at times as a lot of hot sunshine gets in. Also, the steep rake of the front screen tends to result in some sky glare and perhaps Alfa Romeo should consider a band of tint in the upper part (like Volvo).

Rearview mirror is glued to the glass.

Towards the rear, the awkward shapes of the roof pillars don't affect vision too much nor does the flatness of the screen. But we feel a wiper is necessary since the gentle slope does not evacuate water fast enough and light drizzles can smudge the glass very quickly.

Despite criticisms of the layout whereby the speedometer and tachometer are far apart in the GTV, Alfa Romeo did not take the opportunity to come up with an improvement for the GTV6. Thus, the speedometer still remains alone in front of the driver while the tachometer and ancillary gauges are located more to the centre of the dashboard. What's worse is that the tachometer is on the extreme left, more suited to a left-hand drive car. Surely Alfa Romeo can compromise a bit in their quest for a sporty image?

Stalks are used for the more important functions and, unlike some other manufacturers, Alfa Romeo have not seen fit to make modifications to suit right-hand drive markets. This means that the stalks are the opposite of what you'd be used to in a Japanese car.

Running along the transmission tunnel from the dashboard to between the front seats is a console with some space for odds and ends. The cigarette lighter is also located here, adjacent to the handbrake lever, as is the toggle switch for the door mirror adjustment. The gear lever is a typically Alfa Romeo affair — long, topped with a wooden knob and surrounded by a mock leather 'bag' for a classic sporty image.

Storage space in the GTV6 is not particularly generous, restricted to a glove compartment of moderate porportions, flat topped dashboard on the passenger's side and a small shelf.

Strangely enough, the GTV6 is one of the few hatchback-type cars with fix-

The gear lever is a traditional Alfa Romeo design with a wooden knob.

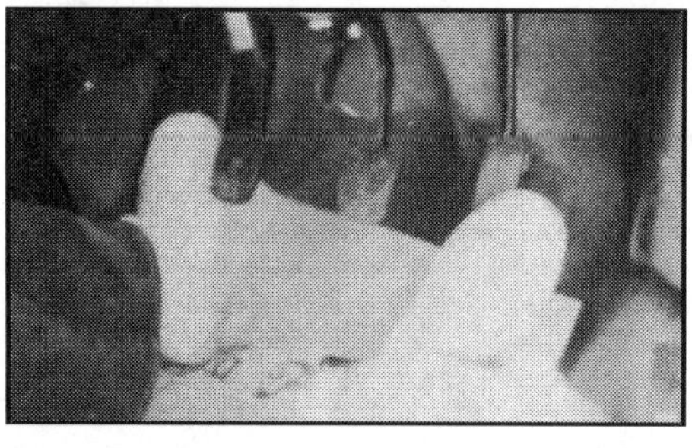

Being a sports car, the pedals are well laid out and structured for heel-and-toe cornering.

ed rear squabs. Apparently, Alfa Romeo feel that a certain amount of safety is compromised with such a feature and have omitted it. Nevertheless, the boot (fully carpetted) holds a large 0.37 cu. metre volume. Unfortunately, the parcel shelf isn't detachable and tilts up as the gas strut beneath it (supporting the hatch when it's opened) telescopes.

The accommodation aspects of the GTV6 can be described as 'reasonable' for this type of car. The absence of the gearbox in front is beneficial to legroom, as is apparent from the deep footwells. With judicious positioning of the seats, the rear passengers should have enough room for their legs although the low rear seat height might mean bending the knees. The bucket design of the rear seats gives good restraint which is needed

The strut supporting the rear hatch is under the parcel shelf which is pivoted up when the strut telescopes.

when the car is driven exuberantly (as most people will drive, we expect). Taller occupants might feel that the ceiling is too close for comfort but this can't be helped in most coupes.

Driving the GTV6

Like testing a BMW, having an Alfa Romeo for testing — especially if it's a coupe version — is always looked forward to with much enthusiasm. So it was with the GTV6 that we kept finding excuses to take the car out as often as possible.

To someone who's spent a lot of time in Japanese cars, the steering of the GTV6 can be a bit heavy initially. The heaviness becomes all the more obvious along twisty roads where, at the end of it all, your arms feel a bit worn out. Thankfully, the gearing allows for easier turning.

There is no doubt that Alfa Romeos are among the finer handling cars in the world and this is demonstrated in no uncertain terms with the GTV6. However, we thought that the ponderous understeer in the initial stages combined with poor self-centering conspired to take the edge off the car's handling as a whole. Mind you, most people will take it as it is and learn to live with it but we think that the front end geometry can still do with some refinement.

Generally, though, the car gets through corners swiftly and with a minimum of drama. The Pirellis on our test car struck tenaciously, allowing accurate lining up of the car for a corner. For those who think the enormous amounts of torque mean oversteer on demand, this is one car that won't oblige. Even if you inadvertently brake in mid-corner, the excellent brake balance prevents any undue reaction at the rear end. It all adds up to making the GTV6 one of the more satisfying cars to drive fast and with a very high degree of confidence.

Ride comfort in a sportscar like this is often third or fourth in importance. With such a taut suspension setup, the GTV6 has a firm ride that is jiggly at moderate speeds but smoothens out as you go faster. It should, however, be noted that the close proximity of the floor to the road means marginally higher levels of shock transmission but we didn't think them unacceptable in the GTV6.

The big V6 emits a characteristic growl and, mixed with the 'corporate Alfa exhaust note', the car's presence can be discerned aurally. Flexibility is part of Alfa Romeo's engine philosophy and the GTV6 possesses a very braod and effective band of power. There's lots of punch and eagerness from near idling speed all the way to the redline of 6250 rpm — and beyond, which Alfa Romeo permit for short durations. Yet, the urgency of the power delivery is not crudely felt as the engine is especially smooth.

Alfa Romeo claim a top speed of 'over 190 km/h' so we were determined to find out just how much 'over 190 km/h' meant. It was difficult to find a reasonably clear stretch of road but for one glorious moment, no traffic blocked the view ahead and we made full use of the opportunity. 210 km/h. Wow!

Getting acceleration times were more tedious than exciting as each run returned a significantly different time. The gearchange action takes some getting used to as the linkage to the gearbox behind causes some loss in positive movement. The lever has to be forcefully wrenched across the gap between 1st and 2nd in order to minimise time loss and from 2nd to 3rd, fractions are lost trying to aim for the right slot. At the end of ten runs, with the clutch smelling, the average time worked out to be 8.8 seconds, as quick as the new BMW 320i we tested last month and certainly nothing to scoff at.

Being a fuel-injected engine, we weren't able to do any proper fuel consumption tests. However, the owner of a GTV6 said he uses an average of 13.5 lits/100 kms (20.4 mpg) which, we suppose, is to be expected from a 2.5 litre engine that has this many kilowatts. No Alfa Romeo literature mentions a fuel cutoff feature for the engine so we would assume it's not there. With electronic control, surely it can't be too expensive to incorporate such a facility which does positively save fuel.

City Motors say that the latest GTV6 shipment costs $79,949.80 each with a.c. and insurance thrown in. Unlike just a few years back, there are now a number of alternatives to the GTV6 which offer similar performance at around the same price. Of course, all of them don't have the classic Italian sportscar charisma which only an Alfa Romeo has and that in itself could be the most appealing aspect of the Alfa Romeo GTV6 2.5. *— Chips Yap*

Test car courtesy of City Motors Sdn. Bhd.

BRIEF SPECIFICATIONS

ENGINE

Type	: 6 cylinders, vee configuration (60°), longitudinally mounted
Capacity	: 2492 cc
Bore/stroke	: 88/68.3 mm
Comp. ratio	: 9:1
Valve control	: Single overhead camshaft (per bank), belt-driven
Fuel feed	: Electronic fuel-injection (Bosch)
Max. power	: 117.6 kW (DIN) at 5600 rpm
Max. torque	: 212.9 Nm (DIN) at 4000 rpm

TRANSMISSION

Driving wheels	: Rear
Gearbox	: Manual 5-speed
Final drive ratio	: 4.1:1

SUSPENSION

Front	: Independent — 4 transverse links, double-acting telescopic dampers, torsion bar, anti-roll bar
Rear	: De Dion rigid axle, oblique trailing arms, transverse Watts linkage, double-acting telescopic dampers, coil springs, anti-roll bar

BRAKES

Front	: Ventilated discs
Rear	: Discs

STEERING

Type	: Rack and pinion
Turning circle	: 10.1 m

WHEELS/TYRES

Wheels	: 6J × 15, light alloy
Tyres	: 195/60HR15, steel-belted

DIMENSIONS & WEIGHT (mm)

Wheelbase	: 2400
Front track	: 1373
Rear track	: 1352
Overall width	: 1664
Overall length	: 4260
Overall height	: 1330
Kerb weight	: 1210 kgs

PERFORMANCE

Max. speed	: 210 km/h
0 to 100 km/h	: 8.8 secs
Fuel consumption	: average 13.5 lits/100 kms (20.4 mpg)

NB : Performance figures obtained on test car only. Actual performance and fuel consumption may vary from car to car depending on driving style, state of engine tune, traffic conditions, etc.

PRICE

On-the-road*	: $79,949.80
Optional metallic paint	: $1,500.00
Optional electric windows	: $680.00

*** On-the-road price includes insurance premium ($2,281.20) for comprehensive coverage for one year for a value of $77,000.**

Alfa Romeo GTV6

THE ALFA GTV6 has, on paper, all the ingredients to be one of the front runners of this year's series production racing. It has the power: at 117 kW, almost 40 per cent more than the Mazda RX7. The engine also has the solid low-down torque and, despite an ignition cutout at 6500 rpm, a power band that's usefully broad, without any rotary or turbo engine peakiness. The gear ratios should be able to harness all this with fine effect for circuit racing, for the first four ratios are short for a car of this power, a fact which shows up in the GTV6's acceleration times.

That those times aren't the best of our group is the fault of two things: on standard road tyres (Pirelli P6s are fitted) the Alfa's traction is so good that fast standing start getaways are difficult. A little more wheelspin would be desirable, and doubtless in race form this could be provided. The other negative feature of the Alfa's performance is the gear change. Even when you've mastered the strong springing and accurate rev matching that's called for, the lever's long travel makes it the slowest change of the cars tested here and also the most worrisome: finding a lower cog double quick is a hit and miss business.

Formidable rear-end grip gets power of terrific V6 on to road with aplomb, but pushes car very wide in tight turns; encouraging oversteer is difficult but effective when achieved, for then Alfa can display its true balance and precision. Driver's job is made harder by the gearchange and driving position, both of which shouldn't be let loose on a road let alone a racetrack; crash-hatted drivers won't like low roof. Pedal layout is okay

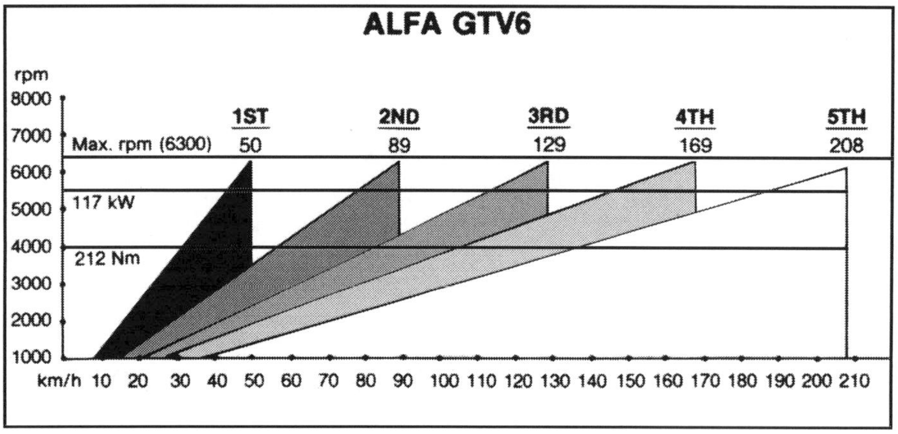

The Alfa driver will also have to contend with the driving position, awkward for many road drivers but potentially very uncomfortable when you have a crash helmet on, especially since the top of the angled steering wheel and the gear lever (in first and third, at least) are a real stretch away.

The seats are designed for comfortable long distance cruising more than circuit racing. Lateral support is fine around the kidneys but falls away towards the backrest's top, allowing shoulders to move around during hard cornering when the car can get up quite a bit of body roll.

That roll gives a clue to the Alfa's suspension set-up, for while you may think this thoroughbred Italian is set up for handling it does, in fact, make many concessions to comfort, a fact which makes it a delightful road car in many respects (see road test page 72 this issue) but which isn't going to do the Group E punter much good.

The car's softness shows up in the body roll, steering and turn-in ability. The steering, a manual rack and pinion system, is normally light and accurate with good road feel, but it loads up considerably through corners, accentuating the wheel's angle. It's reasonably quick with 3.5 turns lock to lock but it comes as a surprise to find how much lock you need for a corner, and how reluctant the car can be to turn in, understeer being prevalent.

Our technique for getting the car to corner more neutrally was to give the wheel one almighty reef. With a bit of a lift-off of the throttle, this would get the front tucking in nicely and the tail, which normally feels glued to the road, running wide, the signal for a bootful of power and flick of opposite lock to get the car four-wheel drifting past the apex, adjustments of line being easily made with the throttle. Get this right and the GTV6 will happily take all the power you can give it, rocketing out of corners with true grip and balance. The possibility of terminal understeer — caused by a missed gear change, perhaps — is the only real threat.

Entirely welcome are the brakes: the four discs, coupled with the car's balance and grip, can be used hard and late without any loss of stability.

Overall, the Alfa GTV6 emerges as a car with potential for Group E honours — but the driver will find it neither the easiest nor the most relaxing of racers.

ALFA ROMEO GTV6-3,0

Developed to win a racing needle match, the sleek three-litre GTV is now South Africa's fastest production car...

THERE'S nothing like a motor racing needle match to spur on the development engineers in competition workshops, especially when it involves two great sporting marques like Alfa Romeo and BMW. With rival models and sporting prides locked in battle on the race track, both teams tend to burn more adrenalin and spend more money in their efforts to gain a consistent edge.

Just occasionally — as in the Alfa Romeo GTV6-3,0 featured in this high performance road test — the road-going enthusiast gets to share the spoils (if he can afford them) in the shape of a sharpened car sold to the public in what amounts to a "limited edition".

To race in Group One, the manufacturer has to accept certain development restrictions; and he has to homologate the model by building at least 200 examples for sale to the public (or selling three months continuous production.) BMW did it with their 535i, which gained more punch on the track when it was fitted with a special close-ratio gearbox. The transmission, which had to be made available to ordinary buyers, hoisted the model's top speed (established on test with CAR) from 206 to 220,7 km/h — and contributed to a series of racing wins over the GTV6.

NARROWING THE GAP

In its standard form, the Alfa was fitted with a 2 493 cm^3 V6, which meant that the 3 453 cm^3 BMW 535i had a litre more engine capacity and 42 more kW — more than enough to offset its mass disadvantage of 260 kg. The gap was too wide and to narrow it, Alfa came back with a three-litre version of their cammy V6, which has enabled them to defeat their rivals convincingly through the 1983 racing season.

Billed as the most powerful production sports car ever built by Alfa Romeo, the car was the result of a co-operative programme run by Alfa South Africa, assisted by Autodelta, the Italian parent company's specialist competition arm, and Alfa's Arese head office. The South Africans were tired of being pipped at the post by 3,5-litre BMWs and thought that an extra half-litre capacity would do the trick.

It did. The new car made its competition debut by winning the Lodge Group One International Two-hour race at Kyalami last year, then came first and second in the Group One class of the Castrol Three-hour race at Killarney. Its speed and reliability were further demonstrated with an Index victory at the World Endurance Championship 1 000 km Race at Kyalami last December and its builders claimed a top speed of over 220 km/h. . .

But was it really faster, in a straight line, than its BMW rival? CAR had previously pitted the 2,5-litre GTV6 against the 535i — then equipped with its new close-ratio five-speed transmission — in a comparative test published in February 1983. So we were keen to try the new "Big Brother" Alfa to assess the gains for ourselves and early this year, we were given our chance.

KEY FIGURES

Maximum speed	224,2 km/h
1 km sprint	29,35 seconds
Fuel tank capacity	75 litres
Litres/100 km at 80	8,09
Optimum fuel range at 80	927 km
*Fuel Index	10,51
Engine revs per km	1 782
National list price	R29 495

(*Consumption at 80, plus 30%)

FASTEST PRODUCTION CAR

As a result, the Alfa GTV6-3,0 now takes the laurels as South Africa's fastest locally built production car with a top speed of 224,2 km/h. But although its power boost and other changes have cut its 0-100 km/h time dramatically from 10,8 to 8,36 seconds, compared with the 2,5-litre model, it cannot quite match the accelerative powers of the muscular 535i, which recorded a time of 7,9 seconds.

The gap is less than half a second but it's there, all the way up to the 1 km post which the BMW flashes past 29,3 seconds after blast-off, with the three-litre Alfa on 29,35. . .

As we noted earlier, Alfa were obliged to build at least 100 versions of the three-litre car, in order to race it in Group One, and when BMW disputed whether this had in fact been done, Alfa South Africa emerged from an official enquiry with honour undimmed. One of these cars was used for this test.

DESIGN ON THE SHELF

The 2 934 cm^3 six retains the functional balance of the smaller motor because both bore and stroke are increased from 88,0 x 68,3 mm to 93 x 72 mm. When the South African company needed to stretch their Group One power plant to fight off the BMW, the basic designs were already "on the shelf" in Milan, for Alfa had designed a three-litre version of their cammy V6, then aborted its production because of the

The engine compartment is dominated by the enormous aircleaner over six single-choke Dellorto carbs (left). Leather-trimmed steering wheel and gearknob are the major interior changes in the 3,0-litre model (right) — the scattered fascia remains. About the best money can buy, the ultra-fat Pirelli P7s are mounted on expensive Compomotive sectional alloy rims (opposite, inset).

CAR TEST

Interior treatment is still spartan, as befits a true sports car, but both seats and door panels are finished in plush Italian velour (above). The handcrafted GRP bonnet is finished to high standards and incorporates an aggressive power bulge.

tax problems which arise in Italy with large capacity cars.

However, the development work needed to put the unit into competition fettle had still to be done and this was carried out by ARSA's competitions department. They imported the cylinder head casting, crankshaft, special pistons and sleeves from Autodelta – the components required for the capacity change – and had larger valves (needed to boost breathing power) made in Italy to South African specifications.

Everything else was done in this country – including the machining of both block and cylinder head – and performance mods were carefully evaluated stage by stage, to ensure that the development work stayed strictly on course.

It was decided to go for low and medium range torque rather than sheer top-end power, to secure aggressive acceleration out of the corners. Big, downdraught Dellorto carburettors were fitted, instead of the electronic fuel injection used on the 2,5-litre motor, to make the mixture easy to tune and achieve optimum low speed torque.

The carbs were those used on the Alfa Six sedan, rechoked and rejetted to suit the larger capacity and bigger valves. To eliminate any chance of restriction, a larger air cleaner was designed and made locally and this alone contributed 7,5 kW to the net output.

AGGRESSIVE POWER BULGE

Apart from its lower, road-hugging stance, the three-litre car's main distinguishing feature is a superbly crafted GRP bonnet which incorporates an aggressive power bulge required to accommodate the revised induction system. This was also locally designed and made,

along with a deeper front spoiler which lowers radiator temperature by 5°C and makes the car marginally faster at the top end.

In toto, the three-litre development programme raised the power output by 8,5 per cent – from 118 kW at 5 600 r/min to 128 at 5 800 – while increasing peak torque by 4,2 per cent from 213 N.m at 4 000 r/min to 222 at 4 300. But the real gain, as previously noted, was in low speed torque, which shows up in a graph – and on the road.

STANDARD GEAR RATIOS

The five-speed rear-mounted overdrive transaxle retains standard ratios, including the 4,1 to 1 final drive, but the car is dropped closer to the tarmac – and given enhanced traction and stability – by fitting fat 205/50 VR 15 Pirelli P7s on Compomotive sectional alloy rims. This lowers the overall gearing to 33,7 km/h per 1 000 r/min in top, compared to 35,6 km/h/1 000 in the standard GTV6, and this, in conjunction with the much higher power-to-mass factor (up from 97,2 W/kg to 112,5, due to a power increase of 8,5 per cent and a six per cent drop in mass) accounts for the dramatic gain in sprinting capability.

MARKEDLY UNDERGEARED

However, the present gearing is not ideal and may be modified in future by using larger rear wheels and/or a close-ratio transmission. For road use, the big Alfa is a magnificent sports machine but is markedly undergeared, requiring almost 3 000 r/min in fifth for the 100 km/h limit and with a top speed limited to a theoretical 218 km/h by a rev-limiter set at 6 500 r/min.

In fact, we recorded 217 km/h in this trim and the limiter was then discarded, at the request of Alfa themselves, to allow us to establish the car's true top speed. This was not a risky procedure because the limit is conservatively set and the motor is good for 7 000 r/min when raced. In this trim we recorded a true figure of 224,2 km/h. At this speed, the rev-counter which is five per cent fast, indicated 7 000 r/min for a true 6 666.

MOST DRIVERS MOVE OVER

The Alfa's power bulge, deeper front spoiler and much lower stance give it the look of a racer that has strayed onto the public highway and causes many drivers to move over, after the first glance in their mirrors. The GTV6-3,0 is also distinguished by its special wheels and red accent striping on bumpers and side mouldings, while inside it is fitted with leather-trimmed steering wheel, Recaro seats and Italian velvet upholstery.

No air-conditioning or sound system are supplied – despite the price of R29 495 – presumably because if they were, they would have to be fitted to the Group One racers, which would suffer a mass penalty as a result. However, for South African use, we regard air-conditioning as essential in a car of this class, because of the high heat input through the strongly raked screens. It is available as a factory-installed option.

RESERVES OF POWER

Apart from its sheer performance and superb handling characteristics which are, in some respects, in a class of their own, the great pleasure to be got from driving this car comes from its engine, which combines enormous reserves of eager power with docile tractability. In top gear, for example, you can amble along at 60 km/h on under 2 000 r/min, then accelerate smoothly and with no trace of brittleness to the car's top speed.

At virtually any intermediate point a stab on the pedal produces a husky burst of acceleration with no need to drop into fourth — a particularly endearing feature in a car whose gearchange, while acceptable, is among its weaker features.

As the test car was almost new, the shift was stiffer than on other GTVs we have driven and by current standards rates as heavy and "mechanical" in character, with a tendency to baulk when pushed across gate really hard.

This actually affected our times for the 0-100 km/h sprint, because two gear-changes are required to reach a *genuine* metric "ton". The motor starts stumbling at around 96 with rev-limiter off, and in this trim we were unable to beat the times recorded with it on. The other limiting factor is, of course, the wheel-spin which has to be controlled to secure the best times.

If the overall gearing (or second) were a shade higher, the motor's husky flow of low- to mid-range torque would produce even better times. But unless you are obsessed by BMW comparisons the issue is really academic, because on road or track the car is spectacular.

THROATY RUMBLE

Despite its lively stage of tune it bursts into a throaty rumble at the first touch of the key and warms up smoothly and quickly. Although the choke control *is* connected, it has no useful function in summer months (when the air cleaner flaps are kept open) and the car pulls away smoothly, without enrichment.

You gentle it along until the motor is thoroughly warmed up and from then on, every thrust on the throttle produces a muffled roar and the Alfa leaps forward like a leopard charging a baboon. The ride is close to the road surface and harsh, as the stiffened spring and damper settings transmit every bump, every crease in the tar, through the close gripping Recaro seat.

There's a few snarls from the exhaust as you thread through traffic with gut-quenching ease, then a series of bellows as you push her up through the gears on reaching the open road. Soon you're cruising briskly with that marvellous ease of the true sports car, the ride smoothing out and the car responding to every nuance of throttle or steering pressure.

Sure, the gearshift has a longish, slightly cranky action that needs knowing and the noise levels are high; but the feeling of piloting a relaxed but powerful cat, ready to pounce at the first breath of a command, is exhilarating, to the man who enjoys fast cars.

SATISFYING FEEDBACK

Perhaps the greatest single satisfaction to be had from this potent GTV is the feedback from its chassis — the feeling that your nerves extend into the extremities of the car. As you work up the speed through the corners, you sense power train and suspension interacting to give perfect control. The wide track and low stance give a feeling of immense stability and the grip of the big Pirellis is such that you — man and machine — hug the road within precisely acknowledged limits.

Initial understeer soon neutralises out, under power, and you are into a widely defined area where the car responds instinctively to steering inputs with a poised, balanced feel. Most sporting drivers will drive with flair and satisfaction, without ever reaching the limits of adhesion. But the enthusiast seeking something more can corner on the throttle with smoothly predictable control.

For the driver who likes to combine with his car, instead of travelling in cocooned comfort, this is one to go for. But of course, the test car was not without its snags. A simple and obvious improvement to ventilation through-flow could be made by transferring the electric window winders from the front of the car — where we would rather have "manual" anyway — to the rear, where the winders are difficult for the driver to reach and where open windows provide a useful extractor effect.

Alfa have not yet introduced the much cleaner fascia design now used in Italy and the fragmented binnacles — the middle one positioned so that the driver's left hand masks some instruments — are something you accept and live with, in a GTV.

SUPERB BRAKES

Predictably in an Alfa of this type and background, the brakes are superb — 269 mm vented discs in front and solid, inboard discs at the rear. Designed to pull the sleek car to a standstill from speeds over 200 km/h, they do it forcefully and stably on only moderate pedal pressures. But their reserves of stopping power make them a little sensitive when working from lower speeds, such as in our 100 km/h braking programme.

Moreover, they tend to *improve* (instead of fade) in initial usage and were producing times of 3,3/3,8 seconds towards the end of our 10-stop test — about half a second quicker than the first stops.

One problem we struck that we're assured won't bother private owners was an irritating interaction between engine movement and throttle linkage. If you were at all tentative with the throttle on pullaway or after a gearchange (as is sometimes unavoidable in traffic), it could set up an escalating jerkiness which shook the whole car. We're told this will be avoided in future production by using a new cable throttle control.

Although fuel consumption is not likely to be a priority concern for the man who runs this type of car, in fact the three-litre GTV has only moderate thirst when driven smoothly at reasonable speeds. As our tables show, it consumes 7,77 litres/100 km at a steady 60 km/h, rising to 8,93 litres at 100, which indicates a theoretical cruising range (on its 75-litre tank) of 840 km.

Driven with verve on the open road, this figure would fall appreciably, but the car would retain an exceptional cruising capability in keeping with its grand touring character. ●

SPECIFICATIONS

ENGINE:	
Cylinders	V6
Fuel supply	six single-choke downdraught Dellorto carburettors
Bore/stroke	93/72 mm
Cubic capacity	2 934 cm^3
Compression ratio	9 to 1
Valve gear	single overhead cam per bank
Ignition	electronic
Main bearings	four
Fuel requirement	98-octane Coast, 93-octane Reef
Cooling	water
ENGINE OUTPUT:	
Max power I.S.O. (kW)	128
Power peak (r/min)	5 800
Max usable r/min	6 300
Max torque (N.m)	222
Torque peak (r/min)	4 300
TRANSMISSION:	
Forward speeds	five
Gearshift	floor
Low gear	3,500 to 1
2nd gear	1,956 to 1
3rd gear	1,345 to 1
4th gear	1,026 to 1
Top gear	0,780 to 1
Reverse gear	3,000 to 1
Final drive	4,10 to 1
Drive wheels	rear
WHEELS AND TYRES:	
Road wheels	Compomotive sectional alloy
Rim width	8J
Tyres	Pirelli 205/50 VR 15
Tyre pressures (front)	220 kPa
Tyre pressures (rear)	220 kPa
BRAKES:	
Front	269 mm ventilated discs
Rear	250 mm in-board discs
Hydraulics	dual circuit, pressure regulator at rear
Boosting	vacuum
Handbrake position	between seats
STEERING:	
Type	rack and pinion
Lock to lock	3,5 turns
Turning circle	10,9 metres
MEASUREMENTS:	
Length overall	4 260 mm
Width overall	1 664 mm
Height overall	1 330 mm
Wheelbase	240 mm
Front track	163 mm
Rear track	163 mm
Ground clearance	122 mm
Licensing mass	1 138 kg
SUSPENSION:	
Front	independent
Type	adjustable torsion bars, stabilizer bar
Rear	De Dion
Type	triangulated axle, coils, anti-sway bars, Watts linkages, dual jointed half axles
CAPACITIES:	
Seating	4
Fuel tank	75 litres
Luggage trunk	340 dm^3

ACCELERATION

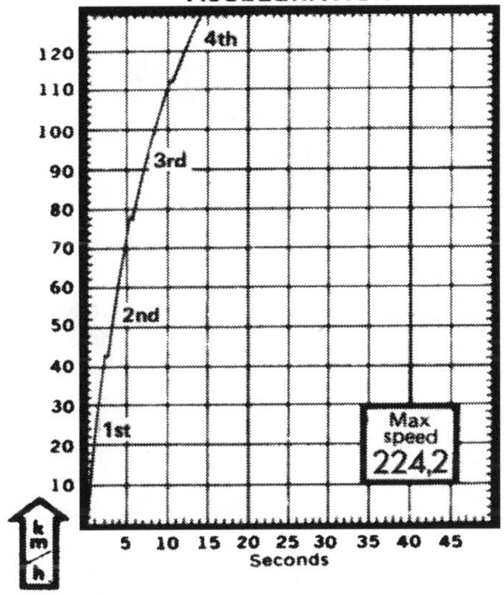

BRAKING DISTANCES

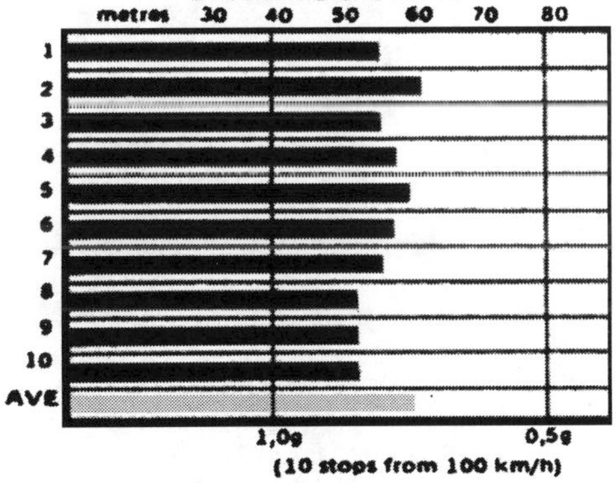

(10 stops from 100 km/h)

NOISE VALUES

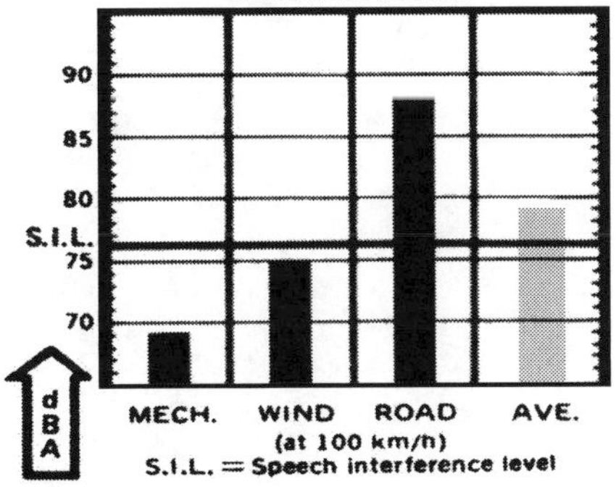

(at 100 km/h)
S.I.L. = Speech interference level

test	Alfa Romeo GTV6-3,0

PERFORMANCE

PERFORMANCE FACTORS:

Power/mass (W/kg) net	112,5
Frontal area (m^2)	2,21
km/h per 1 000 r/min (top)	33,68

INTERIOR NOISE LEVELS:

	Mech	Wind	Road
Idling	53,0	–	–
60	64,0	–	–
80	66,0	71,0	78,0
100	69,0	75,0	88,0
Average dBA at 100			77,3

ACCELERATION (seconds):

0-60	3,56
0-80	5,48
0-100	8,36
1 km sprint	29,35
Terminal speed	178,2 km/h

OVERTAKING ACCELERATION:

	3rd	4th	Top
40-60	2,64	3,85	5,77
60-80	2,52	3,72	5,27
80-100	2,52	3,24	4,92

MAXIMUM SPEED (km/h):

True speed	224,2
Speedometer reading	240

Calibration:

Indicated:	60	70	80	90	100
True speed:	52	62	72	82	93

FUEL CONSUMPTION (litres/100 km):

60	7,77
70	7,90
80	8,09
90	8,38
100	8,93

BRAKING TEST:
From 100 km/h:

Best stop	3,53
Worst stop	4,2
Average	3,9

GRADIENTS IN GEARS:

Low gear	1 in 1,8
2nd gear	1 in 2,8
3rd gear	1 in 3,8
4th gear	1 in 5,0
Top gear	1 in 6,2

GEARED SPEEDS (km/h):

Low gear	43
2nd gear	78
3rd gear	113
4th gear	148
Top gear	195

(Calculated at engine power peak – 5 800 r/min)

TEST CONDITIONS:

Altitude	at sea level
Weather	hot, windless
Fuel used	98-octane
Test car's odometer	706 km

WARRANTY:

Two years or 100 000 km on engine, gearbox, differential and paintwork. One year on other mechanical components.

TEST CAR FROM:

Alfa Romeo South Africa.

ENGINE SPEED

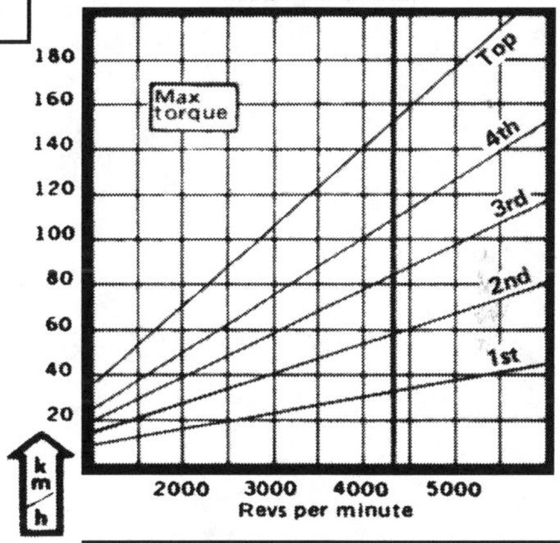

IMPERIAL DATA

ACCELERATION (seconds):

0-60 m-p-h	7,6

MAXIMUM SPEED (m-p-h):

True speed	139,3

FUEL ECONOMY (m-p-g):

50 m-p-h	34,8
60 m-p-h	32,37

GRADIENT ABILITY

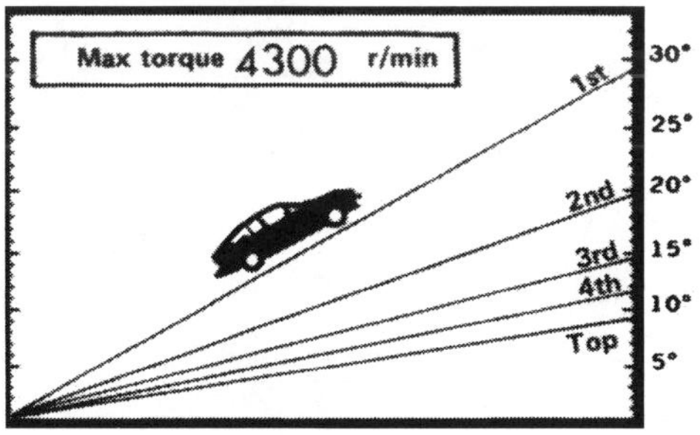

(Degrees inclination)

CRUISING AT 100

Mech noise level	69 dBA
0-100 through gears	8,36 seconds
Litres/100 km at 100	8,93
Optimum fuel range at 100	840 km
Braking from 100	3,9 seconds
Maximum gradient (top)	1 in 6,25
Speedometer error	7 per cent
Speedo at true 100	107
Tachometer error	one per cent over
Odometer error	4,8 per cent over
Engine r/min at 100	2 969

Call me Romeo, Alfa Romeo...

A

GTV6 does James Bond nicely when it comes to saving the world in the film *Octopussy*. Bob Murray finds out how Alfa's flagship fares on less pressing engagements . . .

THE WOMAN in the telephone box seemed just a little upset, and not without good reason. A well turned out woman, she had casually left her Alfa Romeo GTV6 in the market square of a German town while she made her call. Enter, car-less, a harried James Bond eager to save the world. Double oh seven being a proper Brit and all that, he did the right thing and tried to ask her permission to take the car, without success. When the seconds are ticking away to a nuclear holocaust and only you can do anything about it, drastic action is called for and Bond unflinchingly did his duty, jumping into the GTV6, irate woman in background, and roaring off on his mission of mercy that, if successful, would take him to new heights of stardom as the cat's pyjamas of the British Secret Service.

The flagship Alfa sports car is a pretty stylish way to arrive when saving the world, or to arrive at the shops when getting the groceries come to that. On either of these missions, however, one should not take too seriously the saga of the GTV6 as unfolded in *Octopussy*, latest adventure epic starring Roger Moore as a somewhat up-tempo version of the character we once knew as Ian Fleming's. For one thing, the GTV6 is certainly not a car to jump straight into and roar off at a high rate of knots. Fleming's Bond, a Bentley lover, would surely have taken time to warm properly the car's engine and gearbox before such a drive. In *Octopussy*, Roger Moore does not do this, but then that is perhaps to be expected of a man who has ditched his Rolex for a Japanese watch and probably cares not these days if his martinis come shaken with three maraschino cherries and a multicoloured straw.

There are other things to contemplate before driving off in the GTV6, things to savour and to condemn and things that simply demand your attention, for this Alfa even with great familiarity is not a car to be taken for granted. On first acquaintance you will spend at least five minutes adjusting the driving position, and then it will not be ideal, varying from acceptable to bloody awful depending on how far your frame diverges from the short leg/long arm ideal. Reaching the best driving position — and for those on the tall side this is a semi-reclined, knees-up attitude — requires careful balancing of the seat and steering wheel adjustments: set up the seat for easy reach of the gear lever and the pedals will be too close; go for a comfortable pedal position and the steering wheel is out of reach, while wherever you sit

Photography PETER BATEMAN

you will find the roof brushing your hair, the top of the windscreen unnervingly close to your forehead, first and third gears a real stretch away, poor rear visibility and pedals which require an unnaturally high left leg to operate the clutch and which cause an ankle ache in all those without double-jointed right feet. Bond, I'm sure, would have found saving the world a lot easier than having to cope with all this, specially when, true to Alfa's stubborn form, the gear lever works with the smoothness of a rooted box in a '63 Beetle.

Before all this, however, is the GTV6's body. Double oh seven did himself a disservice by jumping straight in for the car's classic Giugiaro styling is worth a few minutes' close examination of anyone's time. There are many things to admire: the rakish, beautifully proportioned profile with the full length side windows and slim pillars that do so much to make the cabin bright and airy; the smart alloy wheels; the superb paintwork (the test car's metallic silver finish made the orange-peeled BMW 323i of elsewhere in this issue seem very second rate, despite the odd bit of grit under the Alfa's skin); the fine fit, finish and sound of panels, hinges and door catches and locks. Even the V6's distinguishing mark, its humped bonnet with black plastic panel, is nowhere near as objectionable as it sounds.

There are delights for the eye at either end of the car, too. At the front there is the lovely alloy V6, the fuel injection gear and large air cleaner nestling neatly between the 60 deg banks with their belt-driven single overhead camshafts. Elsewhere there's all kinds of technical finery such as sodium exhaust valves and electronic wizardry to control ignition timing and fuel delivery. It looks a useful engine and indeed the specifications suggest plenty of getaway performance for overworked spies: 117 kW at 5600 rpm and 212 Nm at 4000 rpm are the claimed figures, and both power and torque are high for a non-turbo'ed engine of just 2.5 litres.

Look closely at the front end and you'll see the wishbone and torsion bar suspension, the anti-roll bar, the steering rack that manages to transmit so much feel to the three-spoke leather rimmed wheel up there in the cabin, and, underneath everything, the driveshaft that takes the power back to the rear-mounted five-speed gearbox, in unit with the limited slip differential.

The de Dion rear axle (with Watt linkage and another anti-roll bar, and springing courtesy of coils) is as much a part of the Alfetta Alfas as that rear-mounted gearbox, if a little less controversial, and makes up a whole that is eminently complementary with Alfa Romeo's image of advanced engineering. Other features help in this direction, too; you don't have to inspect the brakes to know that they are four big servo-assisted discs (though it is nice to push them hard, to know again that an Alfa is probably the best stopper this side of a Porsche), to try the gear ratios to know they will be close to perfect, or to measure the ground clearance to know it is ample not only for Italian motorways but also Australian washaways. Knowledge that the GTV6 comes with a 75 litre fuel tank (the 2.0 litre GTV carries 54 litres) is reassuring, too, but these things are all deep within the Alfa's bowels: visual delight at the rear of the car comes in opening the hatchback rear door and finding luggage space that puts the boots of some sedans to shame, even if the rear seats are of the non-fold down variety, and some space is lost over the 2.0 litre GTV because of the V6's bigger fuel tank.

Our Jimmy had neither passengers nor luggage, nor, one suspects, the three seconds the GTV driver should delay engaging gear after pressing the clutch pedal to the floor. With the clutch so far from the engine (in the GTV it's with the gearbox and diff, unlike the front-mounted clutch in the rear-gearboxed Porsche 924/944) it takes those three seconds for the gearbox to become fully disengaged, allowing cog swapping without graunching. I suspect the graunch in *Octopussy* was edited out; since 007 went on, via some pretty snick driving that included a lovely oversteer slide followed by a rather less lovely bouncing off steel containers, to save the world with only a couple of seconds remaining, we should on this occasion forgive his lack of GTV etiquette.

The GTV6 doesn't sound a hundredth as nice in the film as it does in real life, and in real life it sounds fabulous — probably the nicest sounding bit of hot blooded Italian engineering outside of Maranello or Sant'Agata. The V6's noise is easily one of the car's strongest character traits, in a car that already has more personality in one column stalk than some Japanese coupes have in their entire being. The engine sounds a little clattery at idle, takes on a muted drone at low revs and then develops a musical bark at medium to high revs that leaves you in no doubt as to its condition: it is a picture of health.

The engine's responsiveness is formidable. Most of the adjectives used to describe the sweetest-natured Italian exoticar engines apply, and you know it from the first time you tickle the accelerator and watch the tachometer needle dart for the red line. On the road, this instant and unfussed response makes the V6 one of the most pleasurable engines to drive behind, its lack of any sort of cammy or turbo-effect peakiness enhancing progress around town, often the place where 117 kW engines are either too 'off cam' to cope or too temperamental to be smooth. The GTV6 has no such problems, for there is torque for the asking at very low revs and a mid-range that, with the help of shortish gearing, offers a refined and satisfying accelerative punch — ideal overtaking material. The low-down torque is also ideal for minimising the number of gear changes, good news around town, especially, where coping with the strongly sprung lever is sometimes a dreadful pain. Once in gear, however, the drivetrain behaves well, a slight snatch on the overrun in the higher gears being the only undesirable.

Being an Alfa engine, of course, the power band is as wide as the tacho says it is, and the V6 is willing and strong all the way to the 5700 rpm warning sector, to the 6300 red line and to six-five where an ignition cutout halts further exploration. What an engine this is! In its noise, potency and response — and, yes, its adorable personality — it feels to be straight out the classic exoticar engine mould. And yet at the same time it behaves in such contemporary fashion, giving away little or nothing in even temperament, fuel efficiency and reliability to the most mundane Japanese engine. In the time I had the GTV6, it started first turn of the key every time, performed during a very mixed bag of driving without the slightest tantrum or hiccough, and recorded fuel consumption that varied between a performance-testing worst of 7.1 km/l (20 mpg) and a press-on country cruising best of 9.4 km/l (26.5 mpg). And this from an engine that thinks little of maintaining a cruising speed between 160-200 km/h!

It is cruising at speeds like this, however, that the engine noise, thrown into prominence by a lack of both wind and tyre noise, can become tiresome. Fifth is fairly tall for an Alfa at 35.4 km/h per 1000 rpm but it's still short for a car of this power (the BMW 323i has 41 km/h per 1000 rpm available, for instance) with that consequent lack of quiet for very fast motoring but a decided advantage to performance. The ratios are very much performance oriented, and although not quite as natural as some Alfa gearboxes — the first three ratios can sometimes feel annoyingly short — they do manage to transmit the power with great effect, allowing the rear end to display its considerable traction.

That rear grip, via Pirelli P6s on the test car, is uncommonly good: even sudden applications of power in the low gears generally fail to unstick the GTV6's rear, on wet roads as well as dry, but by the same token full bore getaways can be marked by a deal of axle tramp and thumping, such is the rear tyres' reluctance to spin.

It's been said before — as, indeed, have all the GTV's weaknesses — but the front suspension falls quite a way short of the rear-end competence. In some conditions, in fact, the front end can feel quite incompetent: jittering and moving off line on poor surfaces, turning into corners reluctantly and then running wide in understeer at surprisingly modest speeds, with the steering loading up and requiring strong hands to hold stable. Because of this last feature you would normally say it is a car to drive from the shoulders, but because of the driving position and the wheel's angle, which invariably puts the top of its rim a good stretch away, the steering effort must come from the wrists and lower arms, tiring the driver excessively.

For some of the time on less demanding roads you can rid your mind of these things, saving all your concentration for smooth gear changes and revelling in the engine's appeal, the quiet and supple ride (coupled with

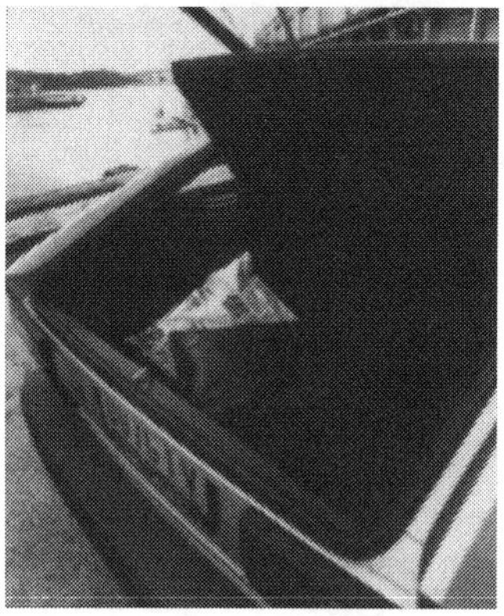

Cabin (top) looks inviting and seats are fine, but driving position, as always feels far from natural for most drivers. Rear seats don't fold and lip is high (above) but boot space is genuine, makes most other coupes appear poorly served

PRICES AND OPTIONS

Make	Alfa Romeo
Model	GTV6
Base Price	$28,400
Metallic paint	$506
Sunroof	$672
Price as tested	$29,578

ENGINE petrol

Cylinders	6
Valves	SOHC
Aspiration	Fuel injection
Comp ratio (to one)	9.0
Bore/stroke (mm)	88 x 68.3
Capacity (cm^3)	2492
Max power (kW/rpm)	117/5600
Max torque (Nm/rpm)	212/4000

TRANSMISSION 5sp man

	Ratios	km/h per 1000rpm
First	3.500	7.9
Second	1.956	14.1
Third	1.345	20.5
Fourth	1.026	26.9
Fifth	0.78	35.4
Final Drive	4.1	

CHASSIS unitary

Drive	rear
Front suspension	wishbones, torsion bars, ant-roll bar
Rear suspension	de Dion axle, Watt linkage, coil springs, anti-roll bar
Brakes	servoed discs
Steering	rack and pinion
Turning circle (m)	10.1
Turns lock to lock	3.5
Kerb mass (kg)	1210
Fuel capacity (l)	75
Tyres	195/60 R15 Pirelli P6

BODY two-door coupe

Exterior (mm)	
Wheelbase	2400
Front track	1373
Rear track	1352
Length	4260
Width	1664
Height	1330

PERFORMANCE

Top speeds (km/h/rpm)	
First	50/6300
Second	89/6300
Third	129/6300
Fourth	169/6300
Fifth	208/5900
Standing start (secs)	
0-50km/h	3.0
0-70	5.2
0-90	7.6
0-100	9.3
0-110	11.2
0-120	13.2
0-130	15.3
0-160	25.6
400m	16.4

In the gears	**2**	**3**	**4**	**5**
30-60	3.2	3.9	6.1	9.2
40-70	3.1	4.3	6.2	9.1
50-80	3.2	4.6	6.4	8.8
60-90	3.4	5.4	6.4	8.9
70-100		5.6	6.7	9.2
80-110		5.8	7.1	9.6
90-120		5.9	7.5	10.4
100-130		6.1	8.6	11.7

FUEL CONSUMPTION km/l (mpg)

Best	9.4 (26.5)
Worst	7.1 (20.0)
Overall	8.7 (24.5)

GTV6's lines are as pleasing as ever (left), full length side windows make cabin bright and aid driver's visibility; near-flat rear window doesn't allow good rear view, however. Standard convenience items abound: headlight sprays are effective

superb high-speed damping control), the light and communicative steering, strong (if a little over-servoed feeling) brakes and smart cabin with firmly bolstered and well trimmed seats. These days there's even a dashboard that puts the major dials in front of the driver (as well as an unacceptable amount of reflections) and sees the cabin full of most of the convenience doodahs one can expect of a $28,000 car. Rear passengers with legs — if you can get them in; the GTV has always looked to have more interior room than it actually possesses — will like the heavily bolstered, kidney gripping seats and the winddown side windows with their unrestricted view out. With all these things in mind and the music of the engine in your ears it is possible to fall in love with the GTV6. And you go and find the good roads.

You will be driving alone and the roads will be familiar. It won't matter if it's raining. Even though you know the road you will be reading the bitumen a good kilometre ahead, following the line of telegraph poles over the treetops, reminding yourself of the access roads and the junctions and the blind, tightening corners that you know could hide a car coming toward you. You will drive smoothly and positively, and you won't have a very comfortable driving position because you will have moved the seat forward to allow your arms more leverage on the steering wheel.

You will know which gears you want long before you want them and on the downshifts you merely dab at the top end of the clutch's long travel before carefully matching revs and slotting the lever home, which, if you did it correctly, the lever will do with consummate ease, even into second. You will use the brakes hard and deep into a corner, and you will use all the engine has to give — heel and toeing at 6000 rpm sometimes. You will take more steering wheel than you consider should be necessary, and you will start to turn in early, allowing the car to drift wide as you feed in the power past the apex, using as much of the rear grip as the front end will allow. Then it's six-three in second, third, fourth . . . yes, fifth, too, for here is a car which red lines in all its gears. If you got the twisty bits right the engine music will be the icing on the cake, and you will permit yourself a quiet smile — if you don't you're not human.

There is enormous satisfaction to be had in driving the GTV6 fast and well, but it's a satisfaction borne not of the car's inherent ability but of your mastering of the car, of your overcoming of the GTV6's weaknesses and inconsistencies, and your ability to draw on the car's inner potential that far too often seems inaccessible. Driving the GTV6 thus is greatly self satisfying: you really do feel as though you've achieved something.

Alas for most of the time, when you're feeling lazy or preoccupied, it is far too easy to feel this flagship Alfa Romeo is a car to fight, and the more you fight it the more it responds badly. Progress becomes jerky and ragged. It is not, as I said earlier, a car to be taken for granted: every gear change, especially the gear changes, and all other driver inputs require careful and considered action, and I'm not sure this is a recipe for a truly capable and relaxing grand touring sports car of the mid-1980s.

It is a shame. I would, I think, happily spend the rest of my motoring behind Alfa's 2.5 litre V6 engine, but I could not contemplate a future in a car that in other respects was so darned frustrating and tiring. The replacement GTV (see following story) will have to be better, but Alfa being Alfa there are no guarantees. In the meantime, the Bonds of this world — James, and our own, equally talented Colin — have their work cut out.

ALFA ROMEO GTV6 — Cont. from page 13

parking, in spite of 3.6 turns lock-to-lock low gearing, and the stylish steering wheel has a thick polished-wood rim which calls for driving-gloves which in maturity I tend to consider affectation, away from the circuits. The gearbox is the weak aspect, and although it is not too bad for unhurried gear-shifts, once on the move (reverse was sometimes very difficult to get), one can see why Alfa Romeo offer only an automatic version of the Alfa 6 saloon. Such a gearbox is out-of-keeping with a low, compact, sporting coupé like this GTV 6, yet one wonders. . . .

I would also like a sun-roof, to augment the rather lackadaisical air-flow from the ventilation system, which has only a 2-speed fan (air-conditioning is an extra) and electric windows, because the body is wide and the window-handles far forward and low-geared. However, it is on the open road that the GTV 6 comes into its own and then there are few complaints — it is a quiet car, it is a comfortable car, the brakes are excellent, the suspension functions well, apart from some slight tendency to weave over certain surfaces, and being too lively over secondary surfaces, and it is a very nicely-finished car. The main thing is, though, that it is a very fast "here-to-there" car, acceleration, even in fifth gear, from say 60 to 80 m.p.h., being most impressive for its response and effortlessness, as the accelerator floors against its metal stop.

The seats are richly upholstered in high-grade velour (leather being an option) and the two rear ones nicely shaped. One has to remember that this is a sports-coupé, however, when contemplating the restricted luggage capacity. Refined touches are the ease with which the front seats slide forward as the backs are folded for access to the rear compartment, the wind-down rear windows, front quarter-lights, the electric driver's door mirror adjustment, the knobs for raising or retracting the front-seat headrests, and a fascia hand-throttle. The door handles, too, are better contrived than on the Alfetta. The fascia layout has been described as untidy, and it does seem odd to fit only the Veglia speedometer into the binnacle directly before the driver, while placing the tachometer, fuel, oil-pressure (55 lb./sq. in.) and heat gauges (175 deg.), balanced by four warning-lights, in another hooded binnacle in the centre, and these are not immune from reflections. On the whole, though, it is not too difficult to get the information needed, and the yellow rear-window demisting and rear fog-lamp buttons with hazard-warning button above them, are admirably placed for the driver's right hand. The transaxle makes for roomy front footwells but lack of a clock, except for a big digital one that is only illuminated when the KE-5300 Pioneer Arc PLL Synthesizer cassette-radio is on, may annoy those who use pocket-watches. Here it can be said that early cars apparently had a better instrument layout, because the catalogue mentions the perfect rational design of the dashboard, with three large, well-lit dials placed directly in front of the driver, i.e., speedometer, electronic engine speed indicator and quartz clock. Be that as it may, there is only a minimum of wood-veneer inside the GTV 6.

The engine peaks smoothly at 5,800 r.p.m., there is a two-plate clutch, the gearbox is at the far end of the propeller-shaft, and the lighting system uses twin Carello headlamps. The three control-stalks are neat, and that for turn-indicators is on the left. The cubby hole is small, adequate for a Canon camera, but unlockable. The outward appearance I thought compactly attractive, sporting without being flamboyant, but the clearance panel on the bonnet is unfortunate. Unless the lockable fuel-filler cap is screwed home, fumes enter the car through the body outlet vents.

Driving the first 200 miles at speed, including Motorway work, then using the GTV 6 disgracefully as a shopping hack, fuel consumption worked out at 23.2 m.p.g. The tank fills easily and as it holds 16½ gallons the range is admirable — one of the few cars in which in times of petrol scarcity I would be just about able to go from mid-Wales to London and back without running out. The engine, which by the way has wet cylinder liners, breaker-less ignition, twin coolant-fans, and is very over-square at 88 x 68.3 mm. bore and stroke, has a 9 to 1 compression-ratio and four main bearings. The o.h.-camshafts are each driven by toothed belts but, as with Jaguar, with the vee engine-configuration, gone is the traditional twin-cam valve-gear. The valves are inclined at 46 deg., enabling good breathing to be obtained. It is said to be somewhat anxious for oil but after 350 miles the level was midway between the marks on the accessible dip-stick. The gearing is 22 m.p.h. per 1,000 r.p.m. in 5th, 16.7 m.p.h. per 1,000 r.p.m. in normal top and in the gears the satisfactory maxima of 30, 55 and 80 m.p.h. can be seen, with "the ton" possible in 4th or top.

To conclude, I found this latest Alfa Romeo a very enjoyable car, and it is not too expensively priced at £9,495, except that some luxury items are omitted, but it is a tiring car to drive in heavy traffic. Its rival must be the Ford Capri 2.8 Injection, which is £1,500 less expensive, but not having driven it I cannot compare these two exciting cars. But I can now understand why Bob Berry, Managing Director of Alfa Romeo GB, enjoys his 2000 GTV, a car enhanced by the new engine. — W.B.

TEST MATCH

SUDDENLY XR4s are everywhere. Ford's high flying Sierra has been on special offer and the price has tempted many buyers.

For Ford it's been yet another way to try and boost the image of the ill-starred Sierra. But is it just price that is winning converts to the XR4i or have Ford managed to tap a vein on the lucrative £10,000 performance car market by merit as well?

To find out, we assembled four quite different variations on the sporting two-door theme. Call the Sierra XR4i a saloon, if you will; it is still the obvious short-list candidate for anyone looking in this price range. So, too, are the more classically coupé-styled Alfa GTV6 and Toyota Supra. For number four, the Audi Coupé 2.2 makes up in price what it lacks in power and it contrasts with the other two coupés here in its whole design philosophy, which makes it all the more worth a closer look.

All four hover closely around £10,000: from the Sierra, cheapest at £9,946 but needing to be loaded up with plenty of extras, through to the £10,999 Toyota.

Already we're tangling with the smallest Porsches: only a little further up the ladder are such desirables as the 944, the Lotus Excel and TVRs.

There's little common ground among our four about what is the formula for a successful challenge to this elite. The Audi is the one most at odds with the consensus: alone of the four it is front-wheel-drive, plain baby sister, of course, to the 4wd Quattro, the majority of whose components it shares.

It has one cylinder less than the others, as well; the engine being Audi's unique five-cylinder overhead camshaft unit. The longish stroke (79.5 × 86.6mm), 2,144cc power unit sits lengthways in the engine bay, and is set well forward at the nose of the car.

Bosch mechanical fuel injection is fitted and the peak power output is 130bhp at 5,900rpm, with peak torque of 126lb ft at 4,800rpm. That makes it, by a substantial margin, the least powerful of the group (as we said, it's price not power that stacks the Audi up against these rivals).

The other three cars use six-cylinder power units but beyond this agreement there is further disunity. The XR4i has Ford's pushrod overhead valve, 2.8-litre V6 with Bosch K-Jetronic injection, identical to that of Capri and Granada. It's an ultra short-stroke unit with 93 × 68.5mm dimensions and delivers 150bhp at 5,700rpm. Peak torque is 160lb ft at 3,800rpm.

Alfa also opt for the V6 format but their's is a smaller, 2,492cc unit that has a single overhead camshaft – belt driven – for each cylinder bank. It uses Bosch L-Jetronic injection and delivers 160bhp at 6,000 revs together with 156.8lb ft torque at 4,000 revs.

Finally to Toyota who have the most powerful and most sophisticted "six" of the test in the shape of a twin overhead camshaft, straight-six unit of 2,759cc which produces 170bhp at 5,600 revs and 169.7lb ft torque at 4,600 revs. Of slightly "under-square" (83 × 85mm) dimensions, it too has electronic fuel injection.

All the cars have five-speed manual transmissions and, although thinking on intermediate ratios differs, they all reach their maximum speeds in fifth so none is geared with fuel saving too much in mind. Thank goodness!

The Alfa, of course, places it transmission at the rear in unit with the final drive to improve weight distribution and handling balance.

Suspension thinking – and complexity – varies greatly, too. Simplest car is the Audi which uses MacPherson struts, with lower wishbones for the front together with an anti-roll bar. At the rear VAG's familiar torsion beam axle is located by trailing arms and a Panhard rod, with coil springing, to give a semi-independent effect.

Quite the opposite extreme is Alfa's sophisticated set-up which uses double wishbones with torsion bar springing and an anti-roll bar at the front, while the rear is a de Dion tube located by trailing arms and a Watts linkage, with coil springing and another anti-roll bar.

Sierra and Supra have similar suspension systems, differing only in their execution. Both use MacPherson struts with anti-roll bars at the front and independent semi-trailing arms

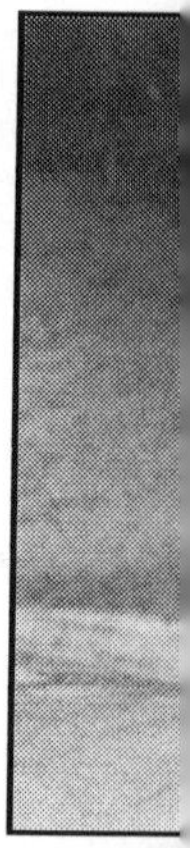

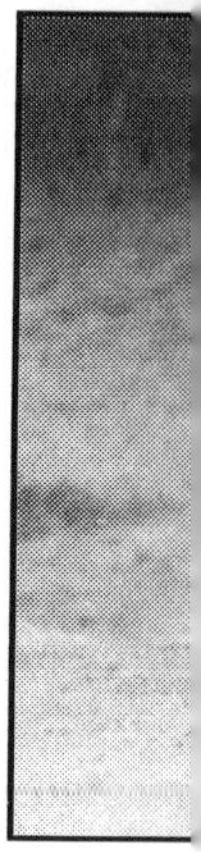

with coil springs and an anti-roll bar at the rear; the Ford having larger, wider-based suspension units than the Toyota.

Alone of the group, the XR4i uses gas-filled dampers all round.

Rack-and-pinion steering is common to all; the Alfa has a manual-only system, the Audi and Toyota have assistance as standard while the Ford has it as a £404 option, which was fitted to our car.

Brakes range from the Supra's big ventilated discs all-round, to the vented front discs/solid rear discs of the Alfa and the vented front discs/rear drums of Sierra and Audi.

All four run on alloy wheels and on 60-series tyres but they range from the 185/60HR14 of the Audi, through 195/60VR14 of the Sierra and 195/60HR15 of the Alfa up to the massive 225/60VR14 of the Supra.

ALFA ROMEO GTV6

WHAT A desperately frustrating car the GTV6 is! There's so much about it that is so good and so much that disappoints.

To most eyes it still looks good; even if the sleek lines of the 1976 original GTV have been brutalised over the years with bigger bumpers, extra plastic and a deep front spoiler. It is still the elegant sporting car of this little group; low, wide, curvaceous – every inch the Italian grand tourer.

The engine, too, is in that great tradition. No-one could ask for a sweeter power unit than Alfa's V6: it howls its approval at being revved with a glorious full-throated roar. Yet it doesn't need to be driven frantically, for it offers surging power right through the rev range; its snap from well below 3,000 revs is remarkable.

Don't think that it is all down to low gearing, for though the first three gears in the 'box are lowish, it will do the same trick in fifth, which is geared the same as all the rest here and will pull it to 132.2mph – fastest of the lot. That makes it a real sparkler on motorways or similar fast roads where it can cruise with the traffic and still have the bite to go when all is clear.

Just as well the engine is so flexible; with a transmission like the Alfa's you need it! Anyone who has read a road test of a GTV since 1976 will only need telling that things haven't got any better.

For those who haven't, the linkage from gearlever to the rear-mounted gearbox in the GTV is a disaster: the shift is slow, indistinct, stiff – every bad adjective about gearshifts applies to this one. Reverse can

prove as hard to find as the left half of a Shell £5,000 Make Money note!

A small sticker on the 'screen warns drivers that the V6 has a twin-plate clutch and to "engage clutch fully before accelerating". Slow, deliberate shifts, using full clutch movements and no throttle until fully completed do make things a bit easier – except that the clutch pedal is angled awkwardly upwards in a way that makes it very difficult to release smoothly.

The other GTV weak spot is its suspension. The sophisticated rear end gives the car masterful traction out of bends and is almost impossible to unstick – though if the tail can be slid, it must be held quickly.

That's the good end; it's the front that seems to be all wrong on GTVs (and varies markedly from car to car, in our experience).

The front suspension on our 5,000 miles-old GTV6 felt as though it had already done well over 50,000 hard miles. Either the dampers were prematurely senile or the car is simply under-damped, for it bounces and wanders disturbingly on the straight – spoiling an otherwise good ride – then rolls and under-steers in an ungainly, tyre scrubbing style round corners. The driver can feel the squirming of the tyres as they try to cope with all the wrong instructions coming from the suspension and respond instead to the superb inputs of the accurate steering.

Alfa don't offer power assistance on the GTV6 and a good thing, too, for it might make parking easier but it would surely spoil the magnificent feel of the steering.

However, the end result of all these quirks is a car that can be tiring and often annoying to try to drive quickly. A combination of a baulky gearchange, awkward clutch and stodgy handling conspire to frustrate the driver at every corner.

That's if he has managed to come sufficiently to terms with the driving position to feel confident in the car. It's the Italianate styling of driving again: arms at full stretch, legs bent up with knees knocking the wheel. Then as you try to get comfortable, using the height-adjustable steering, you find those badly angled pedals cause your knees to foul the wheel rim and prevent any thoughts of heeling-and-toeing on the pedals.

If you could sit correctly in them, the sports front seats would be good as they give plenty of side support. There is a driver's side height adjustment but headroom is pinched in this low-slung car so that's of limited use.

In the back, the seat itself is comfortable but leg and head-room are both very squeezed for adults. Final silliness in the Alfa's design is the pseudo hatchback rear: the boot is shallow but the rear seats cannot be folded down to enlarge it. Not only that but the tailgate has a central gas support strut running under the rear parcel shelf – open the hatch and – bingo! – everything falls off the shelf as the strut goes up.

It's quite well equipped: with headlamp wash, sun roof, electric windows and so on, but what will shock prospective buyers is the poor standard of fit and finish to the interior.

On our car the glovebox was crooked, the centre console was crooked, the fan rattled noisily, doors had to be slammed hard to shut properly, a tatty little bit of carpet was stuck over screws that held down the centre tunnel tray and there was a pervading air of cheapness about the choice of materials.

The Alfa is a car from a different generation than the others in the test and the difference shows up sharply.

AUDI COUPÉ 2.2 GT

THE AUDI is a car that bores you – until you step straight into it from one of the others here and and realise just how good it is in its quiet, under-stated way.

It's three hundredweight – about 15 per cent – lighter than anything else here and it feels very much the lightweight. Somehow it looks smaller than the others, too, even though the Alfa is six inches shorter: its tidy, angular lines give it a compact air.

Come to it cold and you will take a long time to warm to the Audi. The engine feels competent enough but the five cylinders have a distinctive, throbbing unevenness to their ryhthm which becomes a boomy harshness at higher revs and discourages frequent use of the last 1,500 revs in the power band.

If it is revved all the way round, the Audi can acclerate strongly – but it needs full use of the noisy top end. It can just about hold the other three to 60mph at the test track and only starts to slip substantially behind at higher speeds.

But its comparative lack of capacity and the rather high revs at which peak torque comes, give it a loping style that lacks the sprinting punch one associates with a coupé.

Good top speed is some compensation: with nicely chosen gearing the Audi will pull just past peak power to a 120mph maximum in fifth.

However, one quickly runs out of things to say about the engine; it leaves little postive impression – save that of wondering why Audi didn't put their considerable expertise into developing a much smoother six – and creates not a lot in the way of driver excitement, either.

Nevertheless, driving back-to-back with the three other test cars, the extent of the Audi's competence does win praise – grudging at first for a car that seems so lacking in emotion, but then open.

On anything but an arrow-straight motorway where speed alone counts, it holds its own surprisingly well. The engine isn't what keeps the Audi up with the rest, it's the car's all-round agility; its nimble road behaviour.

It's a light, easy, uncomplicated car to drive quickly and yet one whose limits are high and whose behaviour near those limits is impeccable. The very easiness of it all makes it seem unexciting on first aquaintance.

The controls are light – the gearchange slick and fast, though with the remote feel of a fwd car; the brakes responsive, the power assisted steering helpful without being over-sensitive.

Yet drive it harder and it makes no more demands. The gearshift never falters, however rushed the change, and the brakes give a progressive, reassuring stop without the need for undue pedal effort.

The power steering is a well judged compromise, too. It disguises some of the unwanted torque-steer effects of a powerful fwd machine, yet it is direct enough and sufficiently weighted at speed to give the driver a good feel for the handling.

The handling itself is as idiot-proof as it's possible to be. The Audi corners with rather more roll than one might expect but is as near neutral as makes no difference in all but the tightest bends when too-fast entry can naturally make it understeer. Try and provoke it into doing something silly and it will just disdainfully tighten its line and carry on round. It might be the car with the smallest tyres here but it is also the one with the most forgiving handling.

It rides decently well, too, although the understandable jiggliness of a sporty car over poorer roads is made worse by

Audi's usual problem of road thump that makes the ride sound worse than it is.

Coupé the Audi might be, but it will hold four adults in comfort which is far more than most of its namesakes can manage. There is genuine saloon car comfort in the back on a comfortably shaped cushion and with adult-sized legroom. Headroom is a bit tighter but anyone on the lower side of six feet will be okay. Rear passengers might complain about a lack of ventilation, however, as the side windows don't open and the Audi's ventilation system has trouble penetrating the back of the car.

In the front there are big, hard seats with good side support and cushion tilt adjustment. The driving position is basically good though some may find the big gaps between fore-aft seat settings annoying. Headroom is not brilliant but the seating adjustments should keep it tolerable – perhaps that is why Audi have fitted a curious lift-out steel sunroof instead of a nicer sliding one which would lose a couple more vital inches clearance.

It's a bit of a dull interior, though: the tasteful cloth trimmings to seats and doors contrasts with dour plastic mouldings and bleak dashboard with simplistic instruments and clumsy fingertip-style switches (does anyone – other than the Editor – have fingers long enough to work them from the steering wheel?).

Worthy of particular note at this price level is the amount of equipment included: the Audi offers a good package of headlamp washing, central locking, twin remote adjustable door mirrors, electric windows and rear belts, as well as the items already mentioned.

Final high point for this capable all-rounder is its boot: a big, deep, straight-sided luggage bay well able to hold the travelling requirements of the Audi's four adult occupants.

FORD SIERRA XR4i

IF THE Audi is the car of this test that delivers far more than it promises, is the XR4i the opposite? Its dramatic exterior looks with that famous bi-plane spoiler and the red piped body claddings promise so much. Could any car live up to the expectations created by those?

The engine certainly does; the Ford V6 may be the simplest unit here with its central cam and pushrod-operated valves, but it's still quite a peach. It revs strongly and smoothly, giving the satisfying mid-range punch one looks for in a big performance engine.

There's scarcely anything to choose between the track performance of the three "sixes" in this line-up: less than half a second covers them to 60mph and the Sierra, slowest at 129mph round Millbrook is only three mph down on the Alfa, the quickest round the high-speed bowl.

So it's artistic impression rather than technical merit that will split these and the Sierra scores highly: it's strong, willing and has a satisfying rasp to its exhaust when gunned hard.

Its weak spot is slack response at lowish revs: those with longish memories will compare it unfavourably with the now deceased Essex 3-litre V6. That is noticeable from time to time in everyday use when exiting a roundabout or tight bend in the slightly tall second gear; there's a moment's frustration at the slow pick-up. Low down, there's driveline shudder, too.

Though good, in this company it is shown up by the razor-sharp performance of Alfa's V6.

Aside from the slightly high second, the gearing is spot on. The shift has been improved from earlier XR4is, it has chunky, positive action that's not ultra-slick but is good to use.

However, if the engine, with minor reservations, comes up to expectations, the handling, with reservations, doesn't.

Anyone who has driven a Capri injection in anger will be disappointed by the XR4i; the crispness just isn't there. Ford perhaps have made too much effort to retain the Sierra's ride comfort, for it does ride well, and have set the XR4i up too soft.

Though it can be pointed accurately and has the crisp response to steering inputs of the Capri, ultimately it rolls quite noticeably and becomes untidy on the limit, with the rear end indulging in an ungainly lurching that doesn't usually become a tidy, oversteering slide, as the inside rear wheel spins.

It's shown up in its worst light on undulating country roads: driving in convoy behind an XR4i on such a road one could see the lightly loaded inside rear wheel hop into the air on bumpy corners. Inside the car, passengers – especially in the back – feel an uneasy

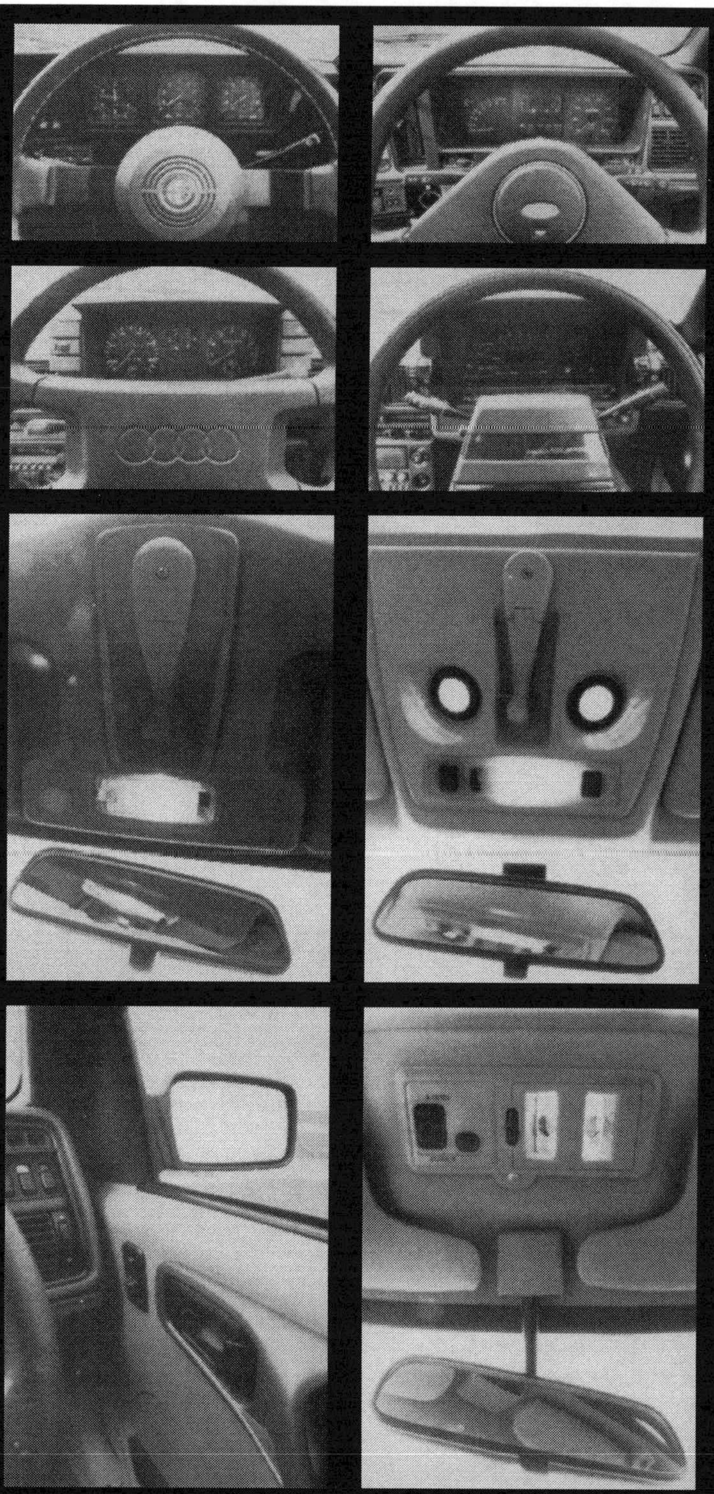

Top: the Toyota's electronic instrumentation is one of the best so far but, in our view, is not as easy to take in as the conventional packs of its three rivals. Of those the Ford and Audi tie for clarity with the Alfa's set up looking just a bit dated and messy. Bottom four pics from top left clockwise: interior treatments in the Audi, XR4, Supra and XR4 again. Right, from top: the GTV6 comes bottom for driving comfort and interior finish. Rather stark but solidly assembled, the Audi has firm, supportive seats and a well-arranged driving position. The Ford and Toyota have similarly opulent interiors and similarly comfortable seats and driving positions, with the Supra having more things to twiddle with if you get bored.

bounciness.

What price a stiffening up at the back end? Poor wet road grip perhaps, for the standard car is no great shakes already on a greasy surface. Standard steering on the XR4i is manual so possibly it's unfair to criticise the optional power assisted system. It is very light, however, and even though the car responds accurately, the system doesn't offer any feedback to the driver of what is going on at the wheels and it also feels very woolly in the straight ahead position.

The rear spoilers are more than high-tech picnic trays, however, for they give the XR4i a lot more high-speed stability than the very side-wind-sensitive standard machine.

But what it does have in common with the more humble Sierras is a roomy passenger compartment. The big, comfortable sports-style front seats have lumbar adjustment and the driver's has height adjustment via an awkward crank handle. A commanding driving position, a nice small, fat wheel and well placed pedals make the Sierra feel good to drive. Some, however, find the big, curved dash something of a muddle with its confusion of differently shaped dials.

There's plenty of room in the back seat – which splits 60/40 to increase the substantial hatchback luggage boot – and the rear side windows open, too, to help Ford's climate control heating system that refuses to deliver fresh enough air when the heater is set much above halfway. Finally, to equipment: though the Ford is the cheapest of the group, central locking, electric windows, headlamp wash, glass sunroof and trip computer are all extras that some or all the others supply standard.

TOYOTA SUPRA

IT LOOKS like some sort of competition car, fresh out of the preparation shop and waiting for its stickers; sitting on those massive tyres that protrude far beyond the lines of the body and need big add-on arch flares to conceal them.

Even the choice of wheels seems designed to emphasise their size; big, spoked alloys like those on a customised Capri.

No, there's nothing subtle about the Supra, and if the exterior treatment doesn't put the message over boldly enough, the interior spells it out in capital letters.

The front seats we've seen before on the old Supra but that doesn't make them any less spectacular; especially the driver's. This has a profusion of adjustments – link it to a microcomputer and it could provide erotic experiences without leaving the car park!

There's height adjustment, side wings to the backrest that can be squeezed in and out but, most bizarre of all, a suggestively shaped bladder that inflates a lumbar adjuster – with three buttons in the backrest to progressively deflate it. Keep quiet those of you who find Recaro seats comfortable without any of this nonsense. There's much less comfort for those in the small back seat, which is better used to extend the modest proportions of the hatchback boot.

Over-elaborate as the front seats are, in the final analysis it has to be said that they are comfortable and the Supra has a super driving position (a height adjustable steering wheel helps).

Flanking the high-tech electronic display instrument pack of the Supra and its digital speedo, strip tachometer and bar graph fuel and temperature gauges, are a pair of notably ugly circular plastic rotary switches for rear wipers and cruise control. The instruments themselves are rather messy looking but do have some nice touches – like an instantaneous conversion to give a detailed "last dregs" reading of fuel level, ranging from quarter tank down through one-sixth to empty. And if you run out after that – you really are daft!

There's plenty of equipment, too; air conditioning, sunroof, cruise control and so on that put the Supra well into the luxury league.

After all this, what the car actually does on the road seems almost superfluous – which is why we come to it last. Most would-be buyers will make up their minds for or against the Supra just looking at it. You don't need telling we came down against!

That's a shame because on the road, it turns out to be one of the best big Japanese sportsters to date. The big tyres do a lot to hold things together on the road: it certainly needs them. Braking deep into a corner or easing the throttle slightly in mid-bend, shades of the old twitchy, tail-happy Supra show in an unsettled feel to the suspension, but the tyres keep it clinging on to much higher levels than before.

It handles with the rugged, classic style of the big, rear-drive traditionalist that it is. There's little body roll through bends; put the power on and it squats down and pushes out of corners on that rubber. It's not agile; it feels a big brute at times on sharply twisting roads, but it's good fun in a slightly old-fashioned way. The balance is much more neutral now – it can be made to tail slide but that needs aggressive driving.

The good power steering is a great help to the handling; firmly weighted and accurate, it feels like the manual steering of a smaller, lighter car. Brakes, too, are good – a relaxing lightness for check stops builds up to the sort of firm pressures one wants in severe braking.

Unfortunately, the engine of the Supra hasn't been improved along with the handling. It's still the same, disappointingly flat unit. The Supra keeps up in straightline acceleration with 0-60mph in 8.2 sec and the top speed is there, too; we lapped Millbrook at 129.6mph. But it lacks punch, never seeming to have the thrust of the other sixes here.

Slightly high intermediate gearing is probably one reason, as is the hundredweight and more of extra weight the Toyota carries. But, even making allowances for these factors, it doesn't feel a particularly sporty unit, and doesn't sound one, either, being decidedly strained at high revs.

On our back-to-back test session its lack of flexibility saw it frequently slipping behind the others; despite best use of the light, fast gearshift. It's a bit of sheep in wolf's clothing, unfortunately, as far as performance goes.

CONCLUSION

WHO SAYS that cars are all the same these days? Or that they are now all so good, there is nothing left to criticise but nit-picking trivia?

We have four cars here that couldn't be more different and we've found plenty to criticise in all of them. Too much, even: it is a deep disappointment that even a £10,000 budget is not enough to find one or two sporting coupés which might be held up as just about spot-on.

Sadly, the car that could be the best has to be classed as the worst. Ever since the GTV came on the scene eight years back, its awful gearshift, indifferent suspension and unpleasant driving position have been repeatedly criticised. Yet they are no better.

Porsche can do it in the 924 and 944, Volvo can do it in the 300 series. Why in eight years haven't Alfa been able to look at those rear-gearboxed cars and find out how? There just aren't any excuses.

There is so much good about the GTV; it could have been Alfa's challenger to the small Porsches with that styling, the prestige of the Alfa marque and that magnificent V6 engine. But those are might-have-beens; the blinkered Alfa management have thrown it all away and left us with an ill-handling, poorly finished machine.

Toyota will show them what they should have done: they have already improved the Supra dramatically in handling and roadholding. Now they need to match their engineering skills to more subjective styling talents and create an interior package worthy of the rest of the car. Undoubtedly they will do it with scarey ease and then the Supra will be a very desirable proposition.

At the moment, it's too garish for our tastes: a car for a successful second division footballer who wants to flash his wealth perhaps.

As the XR4i becomes more common on the roads, so its looks seem easier and less garish on the eye – especially in some, restrained colour schemes. It's a quick mover but another of the group that comes as a disappointment: the suspension is too soft, the steering over-light and the handling not as sharp as the looks suggest.

But for the family man who wants a quick four-seater it has to make sense, shortcomings and all. It offers a lot of space, good finish and the potential for some easy handling improvements.

The Audi makes a lot of sense, too. It can out-corner all of this bunch in its no-frills front-drive way, and its smaller engine is not left far behind on performance as well as giving an economy bonus.

There's room for four inside, the boot is big and equipment is good. But it isn't a car to raise the blood pressure.

In the end, it comes down to the type of motoring you want: the Sierra is extrovert, enjoyable and has punchy performance but it could find a smartly driven Audi Coupé snapping at its heels, driven by a sober-looking executive with the faintest of smiles.

From top: Alfa's superb V6 is the best of the bunch. Smallest of the six-cylinder units, it powers the GTV6 exhilaratingly up to the highest top speed. Audi's straight-five is located way up front – *next* to the radiator. The XR4i's V6 sits much further in the engine bay as does Toyota's massive-looking twin-cam straight-six

	Alfa Romeo GTV6	Audi Coupe 2.2 GTi	Ford Sierra XR4i	Toyota Supra
Price	£10,950	£10,122	£9,946	£10,999
ENGINE				
Cylinders	V6	5 in-line	V6	6 in-line
Capacity, cc	2,492	2,144	2,792	2,759
Bore/stroke, mm	88/68	79/86	93/68	83/85
Valves	Sohc per bank	Sohc	Pushrod, ohv	Dohc
Compression ratio	9.0:1	9.3:1	9.2:1	9.2:1
Fuel system	Bosch L-jetronic	Bosch K-jetronic	Bosch K-jetronic	Electronic injection
Max power, bhp (DIN)/rpm	160/6,000	130/5,900	150/5,700	170/5,600
Max torque, lb ft (DIN)/rpm	157/4,000	126/4,800	160/3,800	170/4,600
TRANSMISSION				
Type	Rwd, 5-speed manual	Fwd, 5-speed manual	Rwd, 5-speed manual	Rwd, 5-speed manual
Internal ratios and mph/1,000rpm				
Fifth	0.78:1/22.0	0.64:1/22.0	0.82:1/22.5	0.86:1/22.6
Fourth	1.03:1/16.7	0.78:1/17.0	1.00:1/18.5	1.00:1/19.5
Third	1.34:1/12.8	1.06:1/13.0	1.26:1/14.7	1.28:1/15.3
Second	1.96:1/8.8	1.52:1/9.0	1.81:1/10.2	1.89:1/10.3
First	3.50:1/4.9	2.85:1/5.0	3.36:1/5.5	3.29:1/5.9
Final drive ratio	4.10:1	4.90:1	3.62:1	3.73:1
SUSPENSION, STEERING, BRAKES				
Front	Independent by double wishbones; torsion bars; anti-roll bar	Independent by MacPherson struts; coil springs; anti-roll bar	Independent by MacPherson struts; coil springs; anti-roll bar	Independent by MacPherson struts; coil springs; anti-roll bar
Rear	de Dion axle located by trailing arms and a Watts linkage; coil springs; anti-roll bar	Torsion beam axle located by trailing arms and a Panhard rod; coil springs	Independent by semi-trailing arms; coil springs; anti-roll bar	Independent by semi-trailing arms; coil springs; anti-roll bar
Steering	Rack and pinion	Assisted rack and pinion	Assisted rack and pinion	Assisted rack and pinion
Brakes, front/rear	Ventilated discs/discs	Ventilated discs/drums	Ventilated discs/drums	Ventilated discs/ventilated discs
WHEELS, TYRES				
Wheels	Alloy, 6J × 15	Alloy, 6J × 14	Alloy, 5½J × 14	Alloy, 7J × 14
Tyres	195/60 HR 15	185/60 HR 14	195/60 VR 14	225/60 VR 14
DIMENSIONS				
Wheelbase, in	95	99	103	103
Front track, in	34	56	58	58
Rear track, in	53	55	57	56
Overall length, in	165	171	175	182
Overall width, in	65	66	68	68
Overall height, in	52	53	55	52
Fuel tank capacity, gal	16.5	15.0	13.0	13.4
Kerb weight, cwt	23.3	20.7	23.7	24.9
PERFORMANCE				
Maximum speed, mph	132	120	129	130
Speeds in gears, mph	(at 6,500rpm)	(at 6,300rpm)	(at 6,300rpm)	(at 6,400rpm)
First	31	31	35	38
Second	57	57	65	66
Third	81	81	93	98
Fourth	105	111	117	125
Acceleration through the gears, sec				
0-30mph	2.6	3.0	2.5	2.6
0-60mph	8.1	8.6	7.9	8.2
0-100mph	22.4	26.2	23.4	23.6
Acceleration in fourth, sec				
30-50mph	5.7	7.1	6.5	8.3
50-70mph	5.6	6.7	6.5	7.9
70-90mph	7.0	8.4	7.8	8.6
Acceleration in fifth, sec				
30-50mph	8.5	10.9	9.0	9.9
50-70mph	8.6	10.5	9.0	10.6
70-90mph	9.8	11.4	10.9	11.1
FUEL CONSUMPTION				
Convoy run, mpg	21.4	24.1	20.9	19.1
Likely variation, mpg	19-28	23-31	19-26	17-25
Government figures, mpg				
Urban cycle	19.1	22.1	18.2	21.7
Steady 56mph	36.7	38.2	37.1	38.7
Steady 75mph	28.5	31.4	29.4	30.1
Maker	**Alfa Romeo, Italy In UK: Alfa Romeo (Great Britain) Ltd, Geron Way, Edgware Rd, London NW2 6LW**	**Audi NSU, W Germany In UK: VAG (United Kingdom) Ltd, Yeomans Drive, Milton Keynes MK14 5AN**	**Ford, W Germany In UK: Ford Motor Co Ltd, Eagle Way, Brentwood, Essex CM13 3BW**	**Toyota, Japan In UK: Toyota (GB) Ltd, The Quadrangle, Redhill Surrey RH1 1PX**

Performance tests carried out at Millbrook Test Track, Bedfordshire.

THE TWO-DOOR sporting coupe market is traditional Alfa territory, and people know it. The other cars in this comparison are lucrative sidelines for their makers, but the GTV6, along with its lesser coupe brothers, is what Alfa Romeo is all about. Indeed, as the flagship, the Six must not only epitomise the company's sporting philosophy but must uphold the image: if you think red-blooded Italian sports cars are the best in the world, here is a car that should be capable of proving it.

It starts out well: 2.5 litre fuel-injected V6 with 117 kW on tap, proper wishbone front suspension with torsion bar springing, a de Dion rear axle with Watts link and coil springs, big discs all round, rack and pinion steering without a hint of power assistance, decently light and good looking body (surprisingly, shorter than even the RX7) that in this company comes as welcome relief after the others' uniform of pop-up headlights, aggressive noses, raked screens and aerodynamic addenda. The GTV6, perhaps with the RX7 a close second, is the only 'pure' body style here, and against it the others look like misfits.

Alfa, like other Italian sports car makers, knows it is no longer enough to build cars that merely look good and go fast, and the GTV6, outwardly the same as always except for new alloy wheels, some detail trim changes and a black bonnet bulge that marks the car as the V6, is a good example of Alfa's increased efforts. The car is better built, better equipped (though, perhaps inevitably, in a tacked-on sort of way), and with the promise of being more reliable and durable than previous Alfas: and there's always the six-year anti-rust warranty to fall back on, something which none of the Japanese cars offer. The price for all this is $29,700, seven grand more than the RX7 and three grand up on the 300ZX.

There is more between the Alfa and its rivals than just dollars, for the car has a very different character to the 300ZX, Supra and Starion. It's a character dominated by the engine.

The V6 has the response, the smoothness, the musical wail and mechanical chatter and the effortless kick in the back from anywhere in the broad rev range to give the car a true sporting personality the likes of which the Starion, Supra and ZX can't hint at. Enthusiastic drivers will revel in this engine, even if they are used to a Ferrari V6: in the way the power is delivered and in the noise it makes doing it, the comparison with Ferrari is entirely valid. It doesn't quite give Ferrari performance, however, and indeed there's not the top end wallop in the Alfa that the fastest of these five cars, the Starion, manages to provide. But the Alfa is always hard on the Mitsubishi's heels up to about 130 km/h (by which time the others have been left behind) and is in one important respect far more capable than the Japanese turbo: in flexibility the Alfa, like the other six-cylinder cars, is streets ahead, picking up promptly and smoothly from low revs in high gear and surging forward while the Starion is still pumping up its turbo.

The penalty for all this comes not in temperament, which is as even as any of the Japanese engines, or in economy, which is absolutely competitive, but in intrusive engine noise: nice sounding although it is, on long runs the engine is tiringly noisy, in marked contrast to the quietness available in the Supra, Starion and ZX. The Alfa is always much noisier, despite well contained wind roar at speed and tyre roar that is not as susceptible to surface change as some of the others. The brakes work well though they are not entitled to be praised as once they could: this is one area in which the Japanese cars have really caught up.

The Alfa's handling, too, must be looked at in the light of the Japanese cars' advance, and if we do this the GTV6 emerges as a safe car in most normal driving conditions, without any trace of the lift-off oversteer available in the others, but it's hardly an inspired handler. The steering is the only one of the group to combine proper road feel

with positive straightahead feel, and these things make the car directionally secure on smooth roads but, as always, on rough ones

there's quite a bit of movement from the front end; the 'Alfa walk' it has been called.

The GTV6 can be taken

Alfa's 2.5 V6 is inspiring but ride quality is a bit agitated and body roll is evident, along with build up of understeer, albeit with best steering. Interior is disappointing, with awkward driving position and instrument placement

through sweepers very fast, and indeed it feels as though the car has been set up specifically for them: there's some body roll, more in fact than the other cars display, and a noticeable building of understeer and of heaviness in the steering, but the tyres stay square to the road, especially at the rear, where there is traction to spare, a feature which gives the car virtual immunity from throttle changes. In tighter corners the understeer and the way the steering loads up are less pleasant, as is the car's often pronounced reluctance to change direction quickly. There's more rewarding road feel and a greater predictability in the car, but overall the Alfa doesn't 'point' as well as any of the other cars in this comparison.

The driver's job is made harder by a litany of unfortunates, surely so well known to readers of this magazine by now that they will roll off the tongue: there's the gear lever, mounted too far forward and with a long-travel and baulk-prone change; the clutch, heavier than the others and with a stupidly long travel movement before an abrupt engagement (and that in the test car was especially awkward); the pedals, high mounted and ache inducing; the driving position, a typical long arm/short leg creation that sees normal-sized and tall people nearly bent double, head uncomfortably close to the roof and top of the windscreen, and the top of the steering wheel a stretch away. Compared with the natural and relaxed driving positions of the others, and their easier to use and more reliable controls, the Alfa seems cursed.

Still, one of our drivers found the Alfa's driving position the best. Art director Gigi also wasn't bothered by the steering heaviness around town, although she did have some scathing words for the dashboard layout and controls location. For the rest of us the driving position, especially, was unacceptable, although it's fair to point out that you can will yourself into living with it, as you can with most of the other drawbacks: drive hard and fast, for instance, and the gearchange and clutch then work a whole lot better, allowing super fast shifts that of this group only the Mazda can match.

Ride quality is a little more agitated than the ZX, a car which also manages to control body roll much more effectively than the Alfa. Still, the GTV6's largely supple suspension makes it an honourable second. Despite its modest exterior dimensions (it's 275 mm shorter than the 300) the GTV6 takes the prize for best accommodation, just beating — because of its extra rear headroom and features like wind-down rear windows — the otherwise equally spacious Supra. You can get four adults in a GTV6 but three is the better number, and this is feasible for journeys of some distance, specially since the seats are firmly and correctly padded and well shaped, although the front seats don't have the lateral support that is available in the multi-adjustable seats of the other cars.

The Alfa's boot is completely different from the others: it's much deeper but much shorter, and there's no fold down rear seat to boost luggage room.

With its wraparound windows, visibility out of the Alfa is better than the others, but there are anomalies. One exterior mirror, and that placed so far back it's out of the driver's sight line, is one too few in our book. The dipped beam is abysmal, far inferior to that of the Japanese cars. And the cabin rear view mirror offers only a glimpse of the road behind that's narrow enough not to include the flashing blue lights on the horizon. There are other annoying things as well. The cabin lighting is pathetic, as is the illumination of the scattered and hard to find minor controls. The instrument display lacks clarity and is not especially attractive either. Ventilation arrangements are haphazard and control of air distribution primitive compared with, say, the push button efficiency of the ZX. And in bright sunlight the windscreen is plagued by terrible reflections, worse than the Mitsubishi's.

Certainly in these things, and in other idiosyncrasies, the Alfa appears crude compared with the Japanese cars, but it's not entirely fatuous to say the Alfa redresses the balance somewhat with its much stronger and more interesting character. If you buy the GTV6 you will have chosen it for this, and its great looking body and, more than anything, its lovely if loud engine. Perhaps things at Alfa Romeo haven't changed that much after all, for the GTV6 is still at heart an individualist's car that looks great and goes fast.

BRED TO PERFORM

Not too often do we get the chance to drive a car so exclusive that if it happened to be written off in an accident, the model would become extinct. And it's even more rare an event when the one-off exclusive is the property of a major car-maker, rather than a backyard creation of dubious integrity.

When Alfa Romeo's Australian chief Ruggero Rotundo invited MM to drive the company's latest show special, the GTV6 "Andalusia", from Sydney to Melbourne, we must admit to feeling slighly apprehensive.

The stark-while Alfa, which got its name from the dancing horse known for its color, elegant bearing and distinctive Roman nose, and was shown on the Alfa stand at the 1984 Sydney motor show, looked superb against a Double Bay backdrop, but how would it look after 800-plus kilometres of Hume highway? One doesn't normally subject a car designed more for the show circuit than anything else to the rigors of the street — let alone the notorious Hume with its interstate transports and endless roadworks strewn with flint-edged screenings. But no. The Alfa people were making a point that this one was not intended for Mollycoddling; that they would like to see it used, that dynamics were as much a part of the Alfa image as the purely visual.

So here we had what could only be described as a very spectacular-looking GTV6 — showing that the Guigiaro design of the mid-seventies is holding its age very well — and the good wishes of the Alfa people that would enjoy our trip. No, we didn't have to use the bonnet "bra" if we wished not to, and yes, they'd even had the foresight to instal a radar for us to make the journey more likely to be a happy one. They'd even partly disarmed the built-in alarm system so that even if we did have to punch in a combination on the keyboard ahead of the gearshift before the car could be started, at least we didn't have to suffer the shrieks of the alarm itself if we weren't quick enough to deactivate it every time a door was opened.

Considering how spectacular the GTV6 looked, it was surprising how little had actually been done to the body. Although the U.S.-made side skirts, wheel arch flares and front spoilers played their part, much of the drama came from the stark white paintwork. Even the standard-issue GTV6 alloy wheels, used for our trip rather than the wider-section white-centre show wheels because they had less offset and would place less strain on the wheel bearings, fitted well into the scheme of things, partly because the vehicle had been lowered slightly at all four corners. Actually, the car had seen a little extra cosmetic attention since the show, gaining extra white enamel on the door handles, fuel filler cap and bumper rubbers, as well as getting a heavier (white) rub strip along the sides.

The car was also fitted with a different "nose cone", which was clear, rather than the white transluscent of the show version, and was drilled out around the grille area, presumably to help airflow into the radiator.

With tinted side and rear windows, the overall effect was a real head-turning motorcar that had every bit as much appeal as your latest aero styling job. Interesting, considering the age of the Alfa's design.

Inside, the Andalusia was the ultimate GTV6. Front seats were Recaro TXs, trimmed like the rest of the seating in real leather and colored a tasteful light grey, there was a beautifully-styled Momo Pininfarina wheel, trimmed in two tones of light and dark grey leather to match the upholstery, and a similarly-presented Momo gearshift knob.

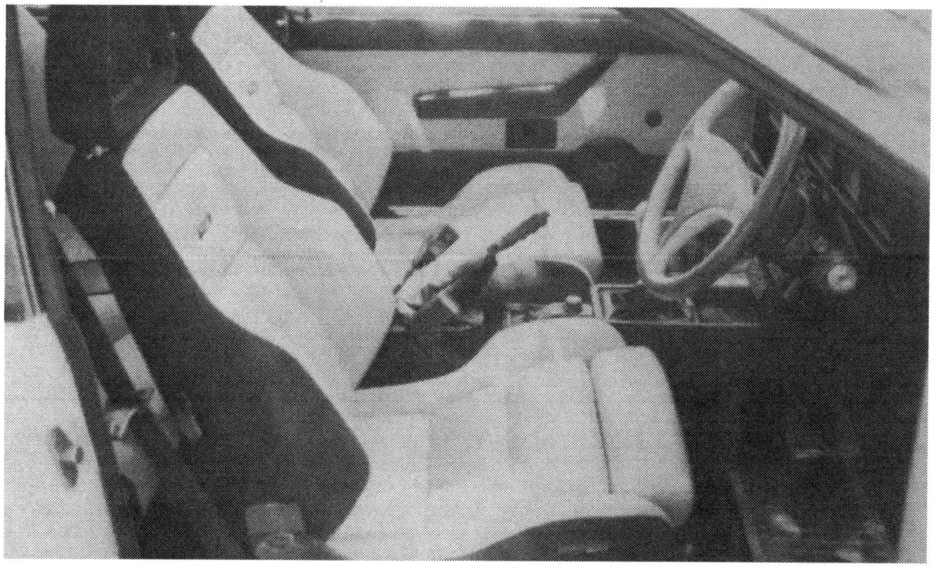

Rounding all this off was a topline Eurovox sound system with power amplifier, twin graphic equalizers, four speakers and separate headphones for the rear seat passengers. Plus, of course, the Super Snooper sitting inconspicuously down the right side of the instrument panel.

Mechanically, the Andalusia wasn't far removed from a standard GTV6. Apart from the lowered and re-rated suspension (springs and shockers both were firmer than standard), a little recalibration of the fuel injection system helped produce slightly more power.

As it turned out, the Andalusia didn't really show any great improvement when put up against the stopwatch, but the suspension job was a definite advantage out on the open road. The GTV6 has always been a glorious motor as much for the audible factor of its exhaust harmonics as its smooth power, but the Andalusia seemed even moreso. Alfa Romeo is one manufacturer which is not about to allow the symphonics of a free-breathing engine to be muffled and strangled by government regulations.

So the V6 encourages you to make full use of its ability to rev, and rewards you with not only brilliant acceleration when you run out to the 6300 rpm red line, but also the sounds that go along with it.

On the open road, the car felt better than we remembered of the last GTV6 turning in with more precision, and tracking very positively at high speed. Despite the extra stiffness of the suspension, it felt better over undulations, too, not so much because it seemed less easily deflected off course, but because the steering response was so much more precise. This was one car you could use to put the kilometres behind you without feeling a great amount of strain after a few hours at the wheel.

One thing we did notice during our time with the Andalusia is that we couldn't really criticise the driving position the way we tend to do with Italian cars. Maybe it was something to do with the use of Recaro seats, but the usual shortcomings weren't as evident — like the normal long stretch to the wheel, and the slight shortage of legroom that underlines the basic differences between Italian and Australian physiques. The Andalusia was very easily to acclimatise oneself to, and felt very "natural" in the way it presented its controls to the driver. Only the gearshift felt typically Alfa in action — but we all know about the shortcomings of that aspect of the car now and so, we suspect, does Alfa Romeo — although it appears part of the problem is the fact that the shift actually works better in a left hand drive car and Alfa Italy is not exactly responding in a hurry to demands from its smaller right hand drive markets.

A very positive point about the car was the fact that if it did have its idiosyncra-

This is the most exclusive Alfa Romeo model in Australia. It's the Andalusia, a one-only machine that has go as well as show. Tim Britten drives it.

sies, it did at least have something to balance that out — character. Driving an Alfa, you may not have the same simple ergonomics of your average Japanese car, but at least you're aware of what you're driving; until you get used to the car, you really have to think about every move you make. An Alfa demands more, but usually ends up giving more.

We only had the Andalusia for four short days, but it didn't become an "anonymous" test car during that time. It was beautiful to look at, an idiosyncratic car that was challenging and enjoyable to drive at the same time. After our long trip, which ended up being broken into two days for reasons disclosed elsewhere in this magazine, we felt more relaxed, comfortable and alert than we had after other long-distance runs in recent times.

One point of hope concerning the one-off exclusiveness of the Andalusia: even though you won't be able to buy one exactly the same, if you're nice to your Alfa dealer, he may be able to come up with some of the equipment for your GTV, such as the side skirts and spoilers. Then, of course, you could approach Recaro for the seats (remembering that they're valued at a cool $1850 for the pair), speak to the people at Carisma about steering wheels and gearshift knobs, and have your windows tinted by any respectable organization specializing in the field (Solar Tint was responsible for this job). The front nose cone and rear filler panel between the tail lights came from Metrix in Narrabeen. Fit a topline radio/stero system like the Eurovox, and you've coming pretty close to having your very own Andalusia.

"... the V6 encourages you to make full use of its ability to rev, and rewards you with not only brilliant acceleration when you run out to the 6300 rpm red line, but also the sounds that go with it."

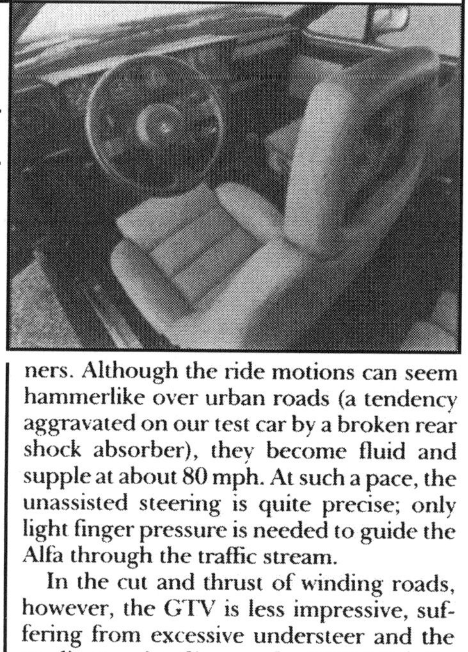

Alfa Romeo GTV6 2.5

The way GT cars were.

• Growing old can be a terrible thing, for cars as well as for people. Some cars do, however, enter their advanced years proudly. The Porsche 911, for example, can wear the mantle of its heritage comfortably because its anachronistic features have been offset by years of intense development. But most elderly machines are forced to soldier on without the constant improvements that might help them defend the fort against the constant attacks of new competitors.

Judging by 1985 standards, the Alfa Romeo GTV6 2.5 seems to be slipping into the elderly-and-ignored category. The GTV has been around since 1974, when it was born as a spinoff of the Alfetta sedan. Its chassis specifications were unusual (and are still unusual today), with unequal-length control arms sprung by torsion bars in front and a well-located de Dion suspension in the rear. These parts were bolted to a rakishly styled unitized body with two-plus-two seating. Alfa's venerable (even then) twin-cam four-cylinder engine powered the new sports coupe through a rear-mounted transaxle; in emissions-legal trim, the car was a bit underpowered. The GTV's driving position was typically Italian, and its instruments were arranged peculiarly, with the speedometer and the tachometer located in the middle of the dash.

Otherwise, though, the GTV was a fine GT car, offering responsive handling, an affinity for high-speed cruising, and plenty of tactile and aural feedback for its driver. In 1981, the GTV was greatly improved with the addition of the 2.5-liter V-6 from the Alfa 6 sedan, as well as a new dash. Compared with other sports coupes of the day, the GTV offered excellent performance, an impressive bloodline, and an attractive price.

In 1985, the aging GTV looks less impressive, but it still has some strong points. The six-cylinder engine remains its best feature, producing a healthy 154 bhp at 5500 rpm while emitting wonderfully soul-stirring sounds. The SOHC engine propels the 2885-pound coupe from 0 to 60 mph in 8.2 seconds, through the standing quarter-mile in 16.1 seconds at 82 mph, and to a lofty 130-mph top speed. In addition to its impressive high-end performance, the engine pulls willingly from 1000 rpm in top gear, with jewel-like smoothness throughout the rpm range.

Part of the credit for the Alfa's excellent performance belongs to the transaxle's closely spaced, relatively low gear ratios. Not even fifth gear has been compromised in search of fuel economy; indeed, the GTV can almost achieve its redline in top gear. This transmission has always been a pleasure to shift, but it did have a couple of problems. For 1985, though, a redesign of the shift linkage, shift forks, selector sleeves, and synchronizers has eliminated the earlier car's crunch when selecting first gear. Gear whine has also been diminished.

The GTV powertrain's fondness for speed is well matched by its chassis manners. Although the ride motions can seem hammerlike over urban roads (a tendency aggravated on our test car by a broken rear shock absorber), they become fluid and supple at about 80 mph. At such a pace, the unassisted steering is quite precise; only light finger pressure is needed to guide the Alfa through the traffic stream.

In the cut and thrust of winding roads, however, the GTV is less impressive, suffering from excessive understeer and the mediocre grip of its new-for-1985 Michelin TRX tires. There's nothing treacherous about its handling, but in contrast to its highway behavior, the GTV is not very responsive in medium-speed corners. The steering is heavy, and the car doesn't cut in toward apexes with any authority.

The arms-fully-stretched driving position doesn't help matters at all. The current GTV does have a vertically adjustable steering column attached to its lovely three-spoked wheel, but it's of little use. At its highest position, the wheel is miles away from you; move it down within reach, and it pins your thighs and obscures the instruments. What this wheel really needs is a telescoping adjustment.

With the decline of Italian imports over

the past few years, we've grown less tolerant of such ergonomic deficiencies. Unfortunately, the driving position is not the only shortcoming in the Alfa's outdated interior. Another is the crude heating system, which requires that individual vents be opened and closed to regulate the airflow. Even with careful fiddling, this system failed to keep the windows defrosted on a cold day (perhaps because outside air was flowing in through a sizable leak under the dashboard).

Our test car's door latches were also no match for cold weather. Although the doors never froze shut, on several occasions they wouldn't close without repeated slamming and manipulation of the latching mechanisms.

The list goes on. Our Alfa repeatedly blew its windshield-wiper fuse at the slightest provocation, just as our last GTV test car had. Its seatback-release mechanisms were incredibly stiff, requiring two hands to operate. The windshield-mounted radio antenna provided reception so poor that the radio had trouble pulling in stations a mere 25 miles distant. At least the radio stayed warm, being heated by air leaking through a small switch panel above it that tended to pop out of place. Light from the glove box also leaked, projecting a bright wedge onto the passenger door at night. Worst of all, a dozen insertions of our radar-detector plugs into the cigarette lighter were enough to tear it free of its mounting and terminate its operation.

Petty annoyances like these have always been part and parcel of Italian sports cars, and enthusiasts used to put up with them because the alternatives were limited. When we tested the GTV6 in 1981, its major competitors were the anemic Porsche 924 and the paunchy and poor-handling Datsun 280ZX. In that company, the GTV's combination of performance and sporting heritage was very attractive.

Today, however, this market segment is overflowing with modern designs. The Toyota Supra, the Mitsubishi Starion, the Nissan 300ZX, the Merkur XR4Ti, the Audi Coupe, and various sporting Mustangs, Camaros, and Daytonas all offer strong performance and handling, without the GTV's niggling ergonomic and quality problems. Further, they cost no more than the GTV's $16,500 base price, and they even offer various convenience features that the Alfa can't match. Faced with such options, even a confirmed Alfa addict can be tempted. For others, the choice is clear.

This need not be so. The GTV's body is still attractive and roomy, its chassis is up to date, and its drivetrain is as satisfying as any in its class. The interior, however, needs to be brought up to the standards of the current decade, and the overall quality must be improved. Otherwise, this elderly Alfa may age for only a few more years before it fades away like its departed Italian and British sports-coupe brethren. *—Csaba Csere*

Vehicle type: front-engine, rear-wheel-drive, 2+2-passenger, 3-door coupe

Price as tested: $18,400

Options on test car: base Alfa Romeo GTV6 2.5, $16,500; leather interior, $750; sunroof, $500; AM/FM-stereo radio/cassette, $450; metallic paint, $200.

Sound system: Blaupunkt Sacramento AM/FM-stereo radio/cassette, 2 speakers, 7 watts per channel

ENGINE

Type	V-6, aluminum block and heads
Bore x stroke	3.46 x 2.69 in, 88.0 x 68.3mm
Displacement	152 cu in, 2492cc
Compression ratio	9.0:1
Fuel system	Bosch L-Jetronic fuel injection
Emissions controls	3-way catalytic converter, feedback fuel-air-ratio control
Valve gear	belt-driven single overhead cam
Power (SAE net)	154 bhp @ 5500 rpm
Torque (SAE net)	152 lb-ft @ 3200 rpm
Redline	6300 rpm

DRIVETRAIN

Transmission	5-speed
Final-drive ratio	4.10:1

Gear	Ratio	Mph/1000 rpm	Max. test speed
I	3.50	4.7	30 mph (6300 rpm)
II	1.96	8.4	53 mph (6300 rpm)
III	1.26	13.1	83 mph (6300 rpm)
IV	0.95	17.4	110 mph (6300 rpm)
V	0.78	21.2	130 mph (6150 rpm)

DIMENSIONS AND CAPACITIES

Wheelbase	94.5 in
Track, F/R	54.0/53.2 in
Length	171.2 in
Width	65.5 in
Height	52.4 in
Frontal area	20.0 sq ft
Ground clearance	5.6 in
Curb weight	2885 lb
Weight distribution, F/R	50.3/49.7%
Fuel capacity	17.0 gal
Oil capacity	7.6 qt
Water capacity	12.8 qt

CHASSIS/BODY

Type	unit construction
Body material	welded steel stampings

INTERIOR

SAE volume, front seat	44 cu ft
rear seat	31 cu ft
trunk space	7 cu ft
Front seats	bucket
Recliner type	infinitely adjustable
General comfort	poor fair **good** excellent
Fore-and-aft support	poor **fair** good excellent
Lateral support	poor fair **good** excellent

SUSPENSION

F:	ind, unequal-length control arms, torsion bars, anti-sway bar
R:	de Dion rigid axle integral with 2 diagonal trailing arms, transverse Watt linkage, coil springs, anti-sway bar

STEERING

Type	rack-and-pinion
Turns lock-to-lock	3.7
Turning circle curb-to-curb	32.8 ft

BRAKES

F:	10.5 x 0.9-in vented disc
R:	9.8 x 0.4-in disc
Power assist	vacuum

WHEELS AND TIRES

Wheel size	150 x 365mm, 5.9 x 14.4 in
Wheel type	cast aluminum
Tires	Michelin TRX, 200/60HR-365
Test inflation pressures, F/R	29/31 psi

CAR AND DRIVER TEST RESULTS

ACCELERATION	Seconds
Zero to 30 mph	2.9
40 mph	4.1
50 mph	5.7
60 mph	8.2
70 mph	10.9
80 mph	14.5
90 mph	19.2
100 mph	25.6
Top-gear passing time, 30–50 mph	8.7
50–70 mph	9.6
Standing ¼-mile	16.1 sec @ 82 mph
Top speed	130 mph

BRAKING	
70–0 mph @ impending lockup	198 ft
Modulation	poor fair **good** excellent
Fade	**none** moderate heavy
Front-rear balance	poor **fair** good

HANDLING	
Roadholding, 300-ft-dia skidpad	0.75 g
Understeer	minimal **moderate** excessive

COAST-DOWN MEASUREMENTS	
Road horsepower @ 50 mph	15.5 hp
Friction and tire losses @ 50 mph	6.5 hp
Aerodynamic drag @ 50 mph	9.0 hp

FUEL ECONOMY	
EPA city driving	**19 mpg**
EPA highway driving	26 mpg
C/D observed fuel economy	**17 mpg**

INTERIOR SOUND LEVEL	
Idle	57 dBA
Full-throttle acceleration	82 dBA
70-mph cruising	75 dBA
70-mph coasting	74 dBA

A111 KRK

BATTLE OF EXMOOR

Exmoor made a tough testing ground for two of the finest coupes around – Alfa's GTV6 and Ford's 2.8i Capri/Gavin Green

Capri is neater of two designs, although GTV6 is more eye catching

EXMOOR WAS IN A FOUL mood that day. The wind howled in from the Atlantic, torturing the trees and buffeting the homes which nestle in the valleys. It was close to gale force nine, the weather men said. The day was still young but the weak dawn sunlight, peeping over the horizon was enough to throw gentle light on the huge grey cloud mass storming at high speed across the sky. There were patches of brightness, and even blueness. But the clouds were still managing to douse the hills and the small towns. The narrow roads, which scar the moors, were glistening with moisture.

We would have to be careful. Yesterday, when we brought two of the most exhilarating £10,000 sports cars available in Britain to Exmoor, the roads were dry. We stormed along with purpose and pleasure, exercising the big wide tyres, the five-speed gearboxes and the muscular V6 engines. Today, great circumspection would be necessary. Leaves were plastered to the tarmac; little streams started to run across the roads. Our pair of budget-priced supercars, each with 160bhp being fed through the rear rubber, would have to be treated with respect.

Yesterday though, they had been in their element. Day one of the 48hour trip to find Britain's most exciting £10,000 car started with a brisk drive along the M4 from London to the Castle Combe circuit in Wiltshire. John McCormick took the Capri 2.8i, now three years old but still Ford's fastest road car. It has just been improved with the addition of a limited slip differential, new wheels and leather upholstery. I drove to Castle Combe in an Alfa Romeo GTV6, also several years old, also V6-powered, also rear-drive, also fuel injected, also based on a slower more prosaic coupe and also one of my favourite driving machines. It was to be a comparison between two cars similar in specification but different in character; of two cars which offer more panache, power and excitement than the latest range of war-painted family hatchbacks to which most enthusiasts now turn, yet don't cost a lot more; of two cars which do most things as well as sports cars costing twice as much.

We could, of course, have added other £10,000 coupes to the list. We eliminated the Sierra XR4i because, although less aged, it is sloppier and more mundane to drive than the 2.8i Capri, even though its appearance may be more strident. We eliminated the Porsche 924 because of the limp-wristed nature of its VW van-based engine, and the Audi Coupe 2.2GT and Mazda RX7 because, likewise, they can't match the Capri or the Alfa in speed. We were not necessarily out to find the best £10,000 coupe; rather the most exhilarating and enjoyable to drive. Big engines and rear-wheel-drive have always proven the most desirable combination when fun is the primary factor.

The Alfa was delivered to my flat early on day one by a bleary-eyed Georg Kacher, in England for the Birmingham Show. The quid pro quo for Georg's delivery job was a lift to Heathrow, and his Lufthansa flight home. The last time I had driven a GTV6 was in Italy, three years earlier, when on mountainous roads in the north the intoxication of the musical and powerful V6 overcame my initial intolerance of the many minor flaws. The main annoyance in Italy quickly came back to light as I attempted to drive Georg away: the gearbox. More vague than a prevaricating politician, the linkages which control the box should have been discarded years ago. Alfa Romeo, who like some Italian concerns are notoriously liberal with the truth, will tell you that the awful change is a necessary corollary of the rear transaxle. Yet a man who says it cannot be done is often interrupted by someone who is doing it; and so the rear transaxle drivetrain arrangement is now used successfully by others. After some gearbox grating and guesswork, with Georg giggling unkindly at my misdemeanours, we were on our way.

Stop at Heathrow, say goodbye to Georg, on to a quick breakfast at Membury services (fast, fried and forgettable) to meet McCormick and the Capri, and the Alfa is starting to ingratiate itself again. The wedge-shaped body, with its amazingly acute angled windscreen, may be getting a big dog eared (not helped by the plethora of plastic addenda which now scar the original Giugiaro design), and the driving position is still more simian than sensible; but the rich growl of the 2.5litre V6 and its incredible tractability (from 1000rpm in fifth gear it pulls without fuss or rancour) is enough to foster a real Milanese love affair.

The engine note is so melodious that owners will find themselves starting the motor to impress curious neighbours. It is so visually attractive they will open the bonnet repeatedly, just to look and touch. The vast alloy castings which cover the pair of single overhead cams are as beautiful as on any Ferrari or Aston Martin; the addenda mounted on top and around are a fitting crown to the jewel. The Italians have always been masters at building beautiful-looking engines: with this V6 powerplant the aesthetic touches are just as impressive as the engineering ones.

Castle Combe was cold, bleak, but dry. It's not your average racing circuit, and that's one reason we like it. Unlike some circuits, which are snooker table smooth, the Wiltshire track is bumpy and uneven. Slight dynamic defects are magnified. Real chassis competence is needed to master the circuit.

I had expected the Capri to be quicker on the track, but was wrong. Ford's best performance car (the 2.8i is a totally different machine from the normal incompetent coupe of the same name) was, after all, developed by people who know how to make road cars go like race cars (Rod Mansfield and the Ford Special Vehicle Engineering team). A few laps in the Capri 2.8 around Donington Park a couple of years back had shown the quickest Capri to be a fine circuit car. Firmly sprung, well equipped with rubber (205/60 tyres front and rear) and powerfully propelled (160bhp at 5700rpm, 162lb ft of torque at 4300rpm), the Ford was nonetheless a consistent 1.5sec a lap slower around Castle Combe than the Alfa.

The gear ratios are the problem. The 2.8i rolls along on a high geared 25.6mph/1000rpm top, with four wide gaps on the way. The ratios just aren't right for circuit work. Many of Castle Combe's corners are too fast for second gear, and too slow for third. As a result the Capri falters coming out of corners, whereas the Alfa with five closely spaced ratios (21.7mph/1000rpm in top) storms out of the bend, with the engine perfectly on cam, tail squatting and revs building. The Capri also tends to die a bit after the first two gearchanges, whereas the Alfa

charges on remorselessly.

The Capri's best lap was a disappointing 1:23.9sec which is the same best lap recorded by the Golf GTi a few months earlier (see *The Sting, CAR* September 1984). It is 1.5sec slower than the Fiat Strada Abarth tested alongside the Golf (same driver, similar conditions, same load). The Alfa's best lap was identical with that of the sprightly Strada (which incidentally is also quicker from 0 to 60mph than the 8.0sec managed by both the Capri and the Alfa).

Gear ratios apart, the other notable high speed disappointment with the Capri is the brakes. Ventilated front discs/rear drums, they lack the bite of the Alfa's all disc set-up (in-board at the rear) and are more prone to fade. The Capri's cornering composure, when pushed hard, is also inferior to that of the Alfa (although it is no slower around the turns; on some corners it is quicker).

Drive the Capri gently, or even at medium to high speeds, and its handling will seem the better. Turn the tiny leather rimmed three-spoker and the Capri's front Goodyear NCT bound alloys grip tidily and neatly and the colour keyed grille points with amazing obedience. There is little body roll and the Goodyears stay quiet. So far, so good. But when you push very hard, in the race against the stop watch, the Capri's composure does start to falter. It will lurch into an oversteer slide, which does neither your heart nor your lap times any good. It is not the sort of transition which can be cured by a gentle opposite lock wrist flick, either. Rather it is of the full bore, tail out, rubber burning variety. Spectacular it may be, and it is fun for the thrill seeker set. But, if you want to go quickly, you're best to be a little more cautious with your entry speed. That way the tail stays under control.

Initially the Alfa won't impress as much. There is more bodyroll, more tyre squeal, more pitch (the springs and their damping are noticeably softer than on the Capri) and the steering (unassisted rack and pinion; the Capri has power assistance) is less direct. Additionally the Alfa's seats don't offer the same lateral support as the leather skinned Recaros which look so purposeful inside the Capri.

Instead, the really impressive thing about the Italian is its chassis balance and poise when the going gets fast. There is no sudden transition, no uncertain high-speed behaviour. Instead the Alfa leans, grips, understeers slightly – and stays that way from turn-in to turn-out. In ultimate roadholding the pair come out with an honourable draw. But in composure and competence, the Alfa wins. The intrinsic superiority of the Alfa's suspension system does tell in the end. No matter how skilled Mr Mansfield and his team, a de Dion-axled trailing arm rear suspension system must form a better high speed basis than anachronistic semi-elliptics (even if the latter has gas dampers and a thick anti-roll bar). Over the bumpier sections of Castle Combe, the Capri is also less settled than the Alfa. It starts to get nervous and begins to lose its touch with the tarmac; the Alfa, in contrast, stays on top.

After lunch at Castle Combe, we head for Exmoor.

The 2.8 Injection – recently uprated – may look like normal Capri but is very different beast with more power, much more competent suspension

I reintroduce myself to the Capri's interior for the run to Somerset. Inside, it is not as classy as the Alfa. Although the Recaros are leather-lined, and so is the small steering wheel, the overall effect is more of plastic penny pinching (this may be a fast car and, at £9500, quite an expensive one, but it is still a Ford). It is also a Capri, and thus old. The dash looks straight out of the 60s (which it is). The instruments are designed in haphazard fashion with four (fuel, ammeter, oil pressure and water temperature) of the six gauges totally blocked by your hands on the steering wheel. There is also a clock, which looks more like the bedside type than a car fitment, at the front of the centre console; it is poorly sited, and is blocked by any paraphernalia kept in the console (like the Alfa, the Capri is not generous with its oddments space).

Unlike the Alfa – which has awkwardly high pedals and a steering wheel which is too far from the torso and too horizontal – the Capri has an excellent driving position. You feel immediately 'right' when you slip behind the Capri's little go-kart sized steering wheel: the Recaros weld your Levis to the leather, the steering wheel itself is set at a comfortable distance and so are the three floor pedals. You look over a long square-shouldered bonnet, complete with power bulge. The front corners are easy to see. Rear vision is worse. Corollaries of the Capri's antiquity include small rear three-quarter windows, and shallow side glass. Reversing can be tricky.

Turn the ignition key and you soon note that the Capri's V6 is more cob than thoroughbred. Whereas the Alfa's V6 is amazingly smooth and musical the Capri's engine has a rorty uneven throb, redolent of V-formation big capacity engines. The twin exhausts highlight the macho, rather than mellifluous, sound. First gear goes in easily, as do all the other ratios. There is a little heaviness about the gearchange, and quite a long throw, but it is direct and pleasant compared with the ambiguous Alfa shift.

On the move the Capri's powertrain has a rather unpleasant drone, again

contrasting with the Alfa's sweet note. It's a familiar complaint, this, with rear driven Fords. Nonetheless the pushrod cast iron unit, despite its lack of breeding and its prosaic looks (the crummy tin rocker covers contrast with the Alfa's vast alloy castings), does rev without drama to its 6100rpm ignition cut-out. There are a few boom periods, but the Capri engine never sounds asthmatic or strained.

It is a more stable car at speed on the motorway than the GTV6, and a more relaxing car too. Although there is too much windnoise (another corollary of the Capri's 15year age), the long-legged transmission means the engine is quiet at speed (only 3900rpm at 100mph in fifth); road noise is also well suppressed. In fact one of the things that surprises you on a long drive in the Capri is just how refined it is.

Off the M5 at junction 25, heading towards Barnstaple on the A361. We almost miss the B3222 turn which is well disguised in the village of Bampton. The Bampton-Dulverton road is a twisting, scenic amalgam of fast and short straights. It's a delight in a car like the Capri. Bodyroll is so well checked, the driving position is so good, the steering is so responsive and the gearchange so fluent that the 2.8i can be punted along at enormous speed. The power delivery, between Goodyear and tarmac, is also much tidier now that Ford have fitted an lsd to the mechanical specification. It's more fun than an Abarth or a Golf GTi. A powerful rear drive car still offers that extra handling balance and fluency of steering (unencumbered by driveshafts juxtaposed with steering rods) that quick fwd machines cannot.

Into Dulverton, set in a lovely valley on the River Barle, and then onto the B3223 heading north-west to Exmoor. The amazing thing about Exmoor is the contrast in scenery. One moment you can be in a picturesque verdant valley among cottages and trees; the next moment the greenery stops and bleak moorland begins. Just such a transition takes place five miles north of Dulverton. You climb a steep winding road, overhung with luxuriant trees. To your left the Barle runs strongly, helping to give the valley its green hue. But then, suddenly, the lushness ends and the bleak world begins. Exmoor is brown, barren, vast and windswept. There are some beech trees, some brown clumps of bush, some sheep and a few tortuous roads on which you can push performance cars. McCormick, Alfa mounted, and Green, in the Capri, race to the hotel slow corners, steep hills and stop at Simonsbath. For £45 per person, including dinner and breakfast, the Simonsbath House Hotel sheltered us from the vagarious Exmoor climate. Next morning the weather was wet and very windy.

Carefully, we headed up to the top of the moor through a high tunnel of hedges, acting as a windbreak to the gales. The Alfa's wipers were thrashing to-and-fro on the steeply-raked screen; the sound of water gushing through the tread of the Pirelli P6s partially drowned the rhythmic beat of the engine. McCormick, also driving circumspectly, was in front in the Capri. We were heading to Brayford, climbing to about 1600ft, not far from the roof of the moor. A few old Land Rovers chugged down the hill, going in the opposite direction. We turned left, down another narrow road, heading towards Sandyway Cross. To our right a large slice of Exmoor suddenly came in full view; an undulating patchwork of greens and browns. But today the weather was so bad that stopping at a beauty spot would have been madness. At Sandyway Cross a hunt was about to start, apparently impervious to the weather. The jodhpur and riding cap set gave us looks of contempt as we threaded our way through the horses, back to open road.

Like the Capri, the Alfa impressed on Exmoor. It is more surefooted in wet conditions than the more firmly sprung Capri. The engine, too, is more throttle responsive and smoother in its delivery than the cast-iron unit reposing in the nose of the Ford. There is a certain jerkiness while accelerating from low speed in the Capri which just does not exist in the Alfa. You get used to the vagueness of the Alfa gearchange, but it is always a source of annoyance. First and reverse are invariably difficult to engage; even the apparently simple changes like 1-2 and 3-4 sometime go awry due to the strangeness of the spring loading. What a pity that magnificent V6 has such an uncooperative transmission.

More impressive than the manner in which the power is sent to the rear wheels, is the manner in which the power is put to the ground. There is a certain surety and security about the Alfa's rear end which cannot fail to impress. The fact that the GTV6 is a natural understeerer is due mainly to the pertinacious grip of that back end, not to any front-end deficiency. At least in the V6 model, though, there is enough power being transmitted through the rear to get *some* tail movement. The four-cylinder GTV models, on the other hand, are inveterate and determined understeerers.

The steering, too, can be annoying. The far off siting of that lovely black leather-rimmed wheel tends to exacerbate the heaviness of the GTV6's steering. Even at mid to high speed, a heave is required to get the Alfa to change direction. To make matters worse the steering is disappointingly vague. It lacks the crispness of the Capri's system. It is also the main cause of the Alfa's inferior directional stability on motorways. For a company whose steering systems are often superb, it comes as a major disappointment.

The instrumentation also used to be disappointing on the GTV6, although it is better now. You used to look over a tachometer in a little binnacle, with all the other gauges grouped in the centre of the dash. This idiotic rather than ergonomic dash has been improved though, with a large tachometer still in pride of place but now flanked by a similarly-sized speedometer and analogue clock. The gauges look more expensive than in the Capri, and so does the rest of the interior, from the velour-upholstered sports seats to the cloth-skinned roof lining. Not that the quality of the finish is good. Like most

Capri (opposite, top) gets 160bhp 2.8 V6. Interior is improved in latest guise with leather upholstery. Handling is great at most speeds, tends to switch to acute oversteer at limit though. Alfa (opposite, bottom) has glorious 2.5 V6. Driving position is terrible. Handling is more composed than Ford's

Alfas of recent memory, some of the minor trim is particularly easy to dislodge. With this Alfa the £11,695 asking price is for the engine; the interior is virtually an auxiliary. Like the Capri, the rear seat is cramped and is better for carrying paraphernalia than people. At a pinch, however, you could carry a pair of adults a short distance. Both cars are equally claustrophobic in the back – although they are better than many other £10,000 coupes, with the notable exceptions of the XR4i and the commodious Audi Coupe. Both have poor boots. The Capri's is worse, with its wide spare wheel greatly reducing the depth of the luggage carrying area. The Ford, however, at least has the benefit of a folding rear seat – something that the Alfa, despite its hatchback configuration, does not. The Alfa is more cramped in the front, though. With its sunroof, there is headroom – just – for 6ft 2in occupants. Taller folk will have to buy the Capri, which benefits from a lower seating position. The Ford, too, has a sunroof.

The roads started to dry as the rain became more intermittent. We headed north, bypassing Exford, towards Porlock and the Bristol Channel. The roads stayed winding, threading their way up and down rises. We

GTV6 now looks a bit dog eared with plastic add-ons which spoil original

Giugiaro shape. Like Capri, performance is fine (130mph, 0-60 8.0sec)

increased our speed. The Alfa does not feel as taut, although it feels safer. The greater bodyroll, the more inept steering and the inferior lateral seating support combine to reduce the effectiveness of Milan's fastest road car. The steering is the main drawback, and the driving position. Whereas, in the Capri, you sit close to the ground, with the three-spoker perfectly placed and with simple wrist flicks the only things necessary to communicate instructions to the front Goodyears. In the Alfa you need muscle power. You have to heave that wheel.

We joined the A39, which runs along the coast, just west of Porlock, and then headed west. It is without doubt one of England's best driving roads, and one of the most scenic. We exercised the muscle cars again, giving them their heads after the confinement of that last narrow stretch of moorland road. The Alfa revs beautifully to its 6300rpm redline, with even more smoothness than the finest of BMW sixes – and with considerably more stimulation. The gear ratios are perfectly arranged. The throttle response is excellent. Even more so than the Capri, you can balance the car beautifully with the throttle. If you're running a little wide of the apex, simply back off and the nose obediently comes into line. Floor the throttle again, and the nose will start to stray a little. The car's composure in such conditions is worthy of the highest praise. You steer with the right pedal almost as much as you do with the wheel.

We drove to Lynmouth, surely one of Britain's most beautiful villages. You go down a one-in-four drop to the village, perched underneath cliffs nearly a thousand feet high at the mouth of the East and West Lyn rivers. Then to Lynton, high above Lynmouth further along the coast, to refuel. The Alfa and the Capri had covered identical distances, under identical conditions. It was 235miles since we refuelled at Membury, a day ago. The Alfa took 9.6gallons; the Capri 10.3. Neither car proved avaricious, despite some hard exercise. In fact the Alfa's 24.4mpg figure, we thought, was commendable. Later on, driving a little more gently, we averaged almost 30mpg.

We headed east, back to Porlock, and then to London. We left Porlock at 5.45pm and arrived in London – after a coffee stop at the Leigh Delamare services on the M4 – exactly three hours later. McCormick, now in the Alfa, reported the Italian car was upset by side winds at around 100mph; Green (Capri mounted) found no such problem. We agreed that the Capri was the car for long motorway trips.

And, after a lot of thought, and not without some regret, we agreed the Capri would be the car to buy. True, it lacks the chassis composure and the intoxicating richness of the Alfa Romeo's muscular V6 engine. It lacks the braking strength, and its interior is not as nice (nor as well equipped: the Alfa offers electric windows and door mirror adjusters, plus more cloth than plastic). Instead the Capri offers more tautness on winding roads, a vastly superior driving ambience, similar 130mph performance, better seating, just as competent a ride (surprising, that), more responsive steering and superior motorway manners. The Capri is also a car in which gear changing is fun, as it should be on any sporty machine. With the Alfa you avoid cog swopping, so poor is that shift. Finally, the Ford comes for around £2000 less than the Alfa. You could add, were you a pragmatist, lower running costs and lower depreciation to the argument in favour of the Capri.

But, please, do not dismiss the Alfa if you are in the running for a £10,000 coupe which offers more flair and more fun than any tarted and tuned family hatchback. As I sit at the typewriter I can still hear the song of that Alfa Romeo engine echoing through the Exmoor valleys.

Alfa's GTV6

Philip Llewellin spends a week with Alfa's GTV6... he was not sorry to see it go

AN EXQUISITE five-speed gearbox, complemented by an equally superb clutch, is one of my 1974 Alfa Romeo Spiders' most delectable features. Although not the lightest of shifts, it combines a top-quality vintage port's unrivalled smoothness with the precision of a surgeon's scalpel. Every change is a delight, even the slow, sympathetic progression from ratio to ratio when the oil is cold. Best of all is the split-second passage from fifth to fourth which taps the 2-litre's power and torque for overtaking, supplements the brakes when approaching a corner and gives you the muscle to exit in style.

In stark contrast, Alfa Romeo's GTV6 is lumbered with a gearbox which epitomises the difference between chalk and cheese. Rear-mounted in the interests of good weight distribution, it appears to be linked to the lever by nothing more sophisticated than a length of old knicker elastic. God knows why it has not been improved since the current GTV replaced its elegant predecessor midway through the last decade. Porsche's 924, 944 and 928 are reminders that a front-engine, rear-gearbox layout can be just as satisfactory as the conventional alternative.

The test car's inherent problem was exacerbated by what may or may not have been a typical clutch. It bit only when the pedal reached the end of its considerable travel. First time out, with the delivery driver in the passenger seat, I had visions of left knee brushing left ear before that critical point was reached. Just as unsatisfactory, albeit for different reasons, was the 911 Carrera Cabriolet driven a few weeks later and destined to be featured in *Sporting Cars* a couple of months hence.

It is rare for exercises of this nature to commence with a few critical paragraphs. Bad points generally punctuate the text when various facets of the car's character are being considered. But the GTV6's gearbox deserves to be given prominence, because it is a fundamental flaw that should and certainly could be eradicated. A sloppy shift would be just another entry in the debit column if we were considering a boring, lacklustre family banger with no more character, charisma and breeding than a Morris Marina. It stands out like a dandelion in the middle of an otherwise immaculate lawn when the car in question is a thoroughbred Alfa Romeo with buckets of *brio* a great deal to be said in its favour.

The GTV6 was styled in the early 1970s and its lines have never appealed to me quite like those of its immediate ancestor, the GTV whose subtle curves were sculpted by Giugiaro long before he left Bertone and founded Ital Design. That model is an affordable classic that would make the perfect partner for my blood-red Spider.

But the GTV6 also has plenty of what it takes to delight the discerning eye. The profile is particularly attractive, flowing gracefully from long bonnet to raked-back windscreen to neatly truncated tail. The 2-litre version looks even sleeker, because its bonnet lacks the angular, plastic-topped bulge needed to house big brother's V6.

That 2.5-litre power unit is as good as the transmission is lamentable — and that means very good indeed. Fuel injection, electronic ignitition and sodium-cooled exhaust valves help account for 160 bhp developed at either 5600 or 6000 rpm. The former is given in the brochure's technical data while the latter coincides with the power curve's peak on a graph elsewhere in the publication. Both agree that the 60° V6 produces 156.8 lb/ft of torque at 4000.

Facts, figures, specifications and statistics can be misleading. What looks good on paper is often less satisfactory on the road, just as a tempting menu does not guarantee a memorable meal. That does not apply to the free-revving GTV6 engine. It pulls like a steam locomotive from very low speeds, thus reducing the need to play find-a-gear games, delivers plenty of mid-range torque and emits a wonderfully rich, businesslike snarl when its Milanese muscles are being well and trule flexed.

Like most cars designed before the first great fuel crisis, the Alfa carries considerably more weight than is fashionable nowadays and is no faster in 0-60 mph terms than the quickest hot hatchbacks. It reached the mile-a-minute mark in 8.1 seconds, matching Alfa's claim and the figure I recently clocked in a Golf GTi. But the GTV6 maintains impressive progress long after its smaller, trendier rivals have reached the end of their tethers. A commendably accurate speedometer suggested its maximum to be slightly higher than the 127 mph quoted in the brochure.

A healthy appetite for miles should not be equated with the sort of thirst that melts your credit cards. I averaged a very respectable 28.6 mpg over 1100 miles, most of which were covered with two adult passengers, and just topped 30 mpg on a couple of long runs. The overall figure is put into perspective by the 28.4 mpg recorded in a Fuego Turbo, the 24.5 mpg of a Starion Turbo and the 26.3 mpg achieved by the BMW 323i featured in last month's issue.

My GTV6 week involved less than two hours on motorways and started with 200 familiar but demanding miles in Mid Wales. Long straights originally engineered for the benefits of Rome's swift-striding legions give way to more interesting roads where power, although always welcome for overtaking, is significantly less important than traction, grip, handling, agility, balance and trustworthy brakes. Less important, in other words, than the factors that make or break a car's sporting reputation.

The red Alfa was predictably efficient on the first part of the route, cruising at three-figure speeds when the coast was clear, maintaining an express train's stability through long, gentle curves and pleasing the ears with an exhaust note rich, strong and melodious enough to win first prize at an eisteddfod. The ride that had attracted critical comments at low speed, while nose-to-tailing through town traffic, was now praised as acceptably firm and right in line with the GTV6's character.

De Dion rear suspension is the most noteworthy feature of a race-bred chassis whose contact with *terra firma* is provided by 195/60HR15 Pirellis mounted on 6J alloy wheels that proved almost as difficult to clean thoroughly as their counterparts on my Spider. I am training an army of ants to tackle the incredibly intricate alloys on the Rover Vitesse now parked on the gravel outside Castell Llewellin's main entrance.

Only that dreadful gearbox prevented the GTV6 rating exceptionally high marks when the chips were down. There are exceptions, but most roads in the entertaining category keep you

busy exploiting every cog from second to fifth. Getting the changes just right, time after time, becomes almost as important as keeping the front wheels pointing in the correct direction, so a vital element was missing on the narrow ribbon of road that wriggles southwards from Devil's Bridge to Tregaron and Lampeter. Few shifts were actually fluffed, but constant care was essential to avoid errors caused by the remote-control linkage.

The GTV6 was otherwise difficult to fault and impressed me more and more as miles and days went by. The steering, heavy for parking and three-point turns, responds beautifully when the car comes alive at speed. Small movements are sufficient to place the fast-moving car with confident, inch-perfect precision and the front wheels transmit messages unaccompanied by excessive vibration or kickback.

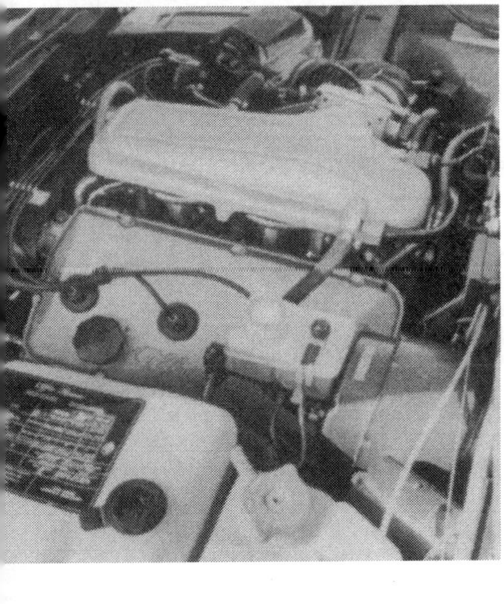

Pirelli's low-profile rubber provides the sort of roadholding that makes you wish for more supportive seats while blessing whoever endowed the GTV6 with a man-sized rest for the left foot. Brutal, low-gear acceleration in a tight corner will ease the tail out, but very gentle understeer is more typical, often accompanied by unwelcome squeals from the tyres. Like the Porsche 928 whose mechanical layout it shares, the GTV6 has a ballet dancer's balance.

Most impressive of all is the way its rear wheels maintain their grip and, avoiding sudden camber changes thanks to the de Dion system, transmit power in situations where most cars would be struggling for adequate traction. You can exercise significantly more horses through fast corners and rocket away from the tighter variety without wasting time on opposite-lock antics. When you boil it right down, few things matter more than keeping a car's driven wheels pressed down good and hard on the tarmac.

Stopping power comes from a quartet of disc brakes which were fine apart from an unfortunate inclination to squeal like the tyres. The GTV6, unlike its four-cylinder clone, has ventilated discs at the front to cope with an extra 30 bhp and additional weight.

Although some three inches longer than a Citroen BX or Ford Orion, this is not the car to contemplate if you worship at the altar of spatial efficiency. The same can be said for the vast majority of eye-catching sports cars, because it is virtually impossible to combine rakish lines and realistic overall dimensions with a capacious cabin. Headroom was marginal for me, even with the driver's seat at its lowest setting. When it was fully raised, I cold almost look out of the sunroof and pretend to be commanding a tank. The relationship between seat, wheel and pedals often makes Brits complain about Italian cars in general, but I had no complaints thanks to being short in the leg.

Although voted comfortable on long journeys, apart from the aforementioned reservations about lateral support, the front seats suffer from non-adjustable head restraints with frames hard enough to give Mrs Llewellin a sore patch on the back of her skull. A couple of inches of vertical movement would have enabled her to doze in the passenger's seat with her cranium cradled by the restraint's netting.

"The roomy interior has space for two passengers to be comfortably seated in the rear," says the brochure, supporting its assertion with a picture showing the front seats pushed well forward while their squabs are almost vertical. It all depends what you mean by two passengers. The usual self-behind-self check had my head hunched forward and my knees spread wide to clear the driver's seatback.

However, it is only fair to report that Mrs Llewellin covered about 80 main-road miles in the back rather than risk being reported to the RSPCA by an elder son a couple of inches taller than his father. She said it was fine once she snuggled down a bit. But there can be no dodging the fact that space is at a premium in the GTV6, and its lack is emphasised by dark upholstery if you specify red or metallic silver paintwork. Beige trim makes the cabin look less claustrophobic, but is only available with gun metal or metallic black paint. Alfa Romeo bodywork should be red.

The boot will take a reasonably big suitcase, plus a fair amount of squashy stuff, but cannot be extended by gaining acces to the passenger compartment. Despite its shape, the GTV6 was designed before versatile hatchbacks became fashionable. A couple embarking on a two-week holiday would have to plan carefully unless they were willing to put luggage on the back seats. A couple with two children would probably have to invest in a roof rack or solve the problem by heading for a nudist resort.

The GTV6's age is also indicated by an instrument panel that looks more late-1960s than mid-1980s. It would make far more sense to cluster the fuel, water temperature and oil pressure gauges in the space currently occupied by a clock as big as the speedometer and tachometer. The latter is right in front of the driver, presumably because Alfa consider the engine's welfare to be more important than speed over the road. I rather like that.

This £11,320 GTV6's ventilation is a joke, even in mild weather. We rarely travelled far without lowering the power-operated front windows and thereby raising noise levels to the point where conversations cannot be conducted in civilised tones. Incidentally, Alfa thoughtfully provide an emergency handle in case the power fails.

As readers patient enough to remain with me until the bitter end will have realised by now, the GTV6 is a somewhat bewildering blend of good and bad mingled with only a few dashes of indifferent. It has plenty of character and is almost worth considering just for the sake of that V6. But could you learn to live with that gearbox?

ALFA ROMEO GTV/6 2.5 & CALLAWAY TWIN TURBO

ROAD & TRACK ROAD TEST

A most unreluctant dragon

PHOTOS BY JOHN LAMM

THE ALFA ROMEO GTV carries a lot of tradition within its familiar wedge-shape form. The current Giugiaro-designed 2+2 coupe, with its rear-mounted transaxle and De Dion suspension, has been around long enough that some of our younger readers may not remember it as the 4-cylinder, 2.0-liter Alfetta GT that made its debut in May 1974. In June 1975, when it was new to America, we chose the Alfetta coupe as the Best Sports-GT in the $8000–$12,000 category in our list of 10 Best Enthusiast Cars, and in June 1978, as the Sprint Veloce, it took the same honors. Three years later, with the insertion of the strong and willing 2.5-liter V-6 engine, the GTV convincingly outpointed the Datsun 280ZX and Porsche 924 Turbo in our Gran Turismo comparison test (July 1981). The GTV label itself (V for Veloce, of course) dates back to the old Bertone 1.6-liter Giulia (also Giugiaro-styled) of the Sixties, and *that* car was a development of the famous Giulietta Sprint of the Fifties.

The last four years have not been particularly kind to the GTV/6 2.5, or rather the competition hasn't. We've seen a new Nissan 300ZX, a Porsche 944, a Toyota Supra and the GSL-SE version of the Mazda RX-7, all cars of more recent conception and greater refinement. In our most recent test of the GTV/6 2.5 (*Road & Track's Guide to Sports & GT Cars 1984*), we criticized its vague shifting, heavy steering and dated interior.

For 1985 Alfa Romeo has addressed some of these shortcomings, particularly the gear linkage, and has made the GTV/6 more competitive by removing some items of luxury equipment and trimming the base price to $16,500 (down from $19,000). The new shift linkage is a clear improvement, allowing confident and quick—though still slightly stiff—changes up or down. There's still a bit of synchro "wait-for-it" in 2nd, but the vagueness of the old linkage is gone and will only be missed by a few traditionalists who took pride in mastering it. The change has made a positive difference in the Alfa's usability and appeal.

Making leather an option because of the lower base price, Alfa has come up with nice fabric-covered seats that pleased most of the staff but still made problems for certain drivers despite the adjustable steering wheel. The main difficulty is with the legs; the seat cushions don't give enough thigh support and the driver's left knee may become sore from the leg being angled excessively toward the clutch pedal. The rest of the interior is still Dark Ages: an unattractive hodgepodge of controls in an

Callaway Twin Turbo adds complexity to engine compartment.

excessively angular black panel, wispy ventilation and several examples of poor quality control.

The V-6 engine has always been the car's main appeal; it still makes those great Italian ripping-raw-fabric sounds, has lots of torque and sends the car down the road smartly. In the standard 1985 GTV/6 we got substantially better acceleration times than in 1984, 0–60 mph in 8.5 seconds, for example, and 24.3 sec to 100. These are more than a half-second quicker to 60 and nearly 3 sec quicker to 100, and the improvement is welcome: a much better performance-per-dollar package.

But forget all that. Here comes the Callaway Twin Turbo, a conversion that transforms the Alfa into a supercar worthy of the Nuvolari and Fangio tradition. How about *three* sec quicker to 60 mph and *twelve* sec quicker to 100? How about mean, ripping-raw-concrete power? Power that comes in at 3000 rpm, demands all of your attention at 4000, and swings the needle to redline so quickly that you'd better be watching the tach or have a really good ear. It's a blast—in a way that the much used word seldom really means. Full acceleration pushes you back into the seat and even makes the belt rachet you tighter. Nothing incredible happens at less than 3000 rpm but if you slip the clutch to

AT A GLANCE

	Callaway Twin Turbo GTV 6/2.5	Maserati Biturbo E	Porsche 944
Price, base/ as tested	$23,994 $27,234	$27,160 $28,985	$22,950 $23,575
Curb weight, lb	2890	2775	2805
Engine/drive	V-6/rwd	V-6/rwd	inline-4/rwd
Transmission	5-sp M	5-sp M	5-sp M
0–60 mph, sec	6.2	6.3	9.0
Standing ¼ mi, sec @ mph	14.5 @ 97.5	15.0 @ 91.0	16.4 @ 82.5
Stopping distance from 60 mph, ft	157	150	141
Lateral acceleration, g	0.81	0.81	0.82
Slalom speed, mph	60.1	55.8	61.4
Fuel economy, mpg	18.5	18.0	21.5

	Pro	Con
Twin Turbo GTV 6/2.5:	a very powerful turbo engine that's also tractable, good high-speed handling, improved shifter	less suitable for city driving, heavy controls, poor ventilation system, outdated dash
Biturbo E: tested 7-85	excellent power with west-coast available intercoolers, opulent interior, exclusivity	turbo lag, poor high-speed braking stability, uneven workmanship and materials, small trunk
944: tested 8-83	exceptionally balanced handling and road feel, usable engine performance, high quality, good mileage	rear seats only for children, ride sometimes jiggly, steering wheel too low for some

The stock GTV/6 2.5 for 1985 is more than 0.5 sec quicker 0–60 than the 1984 model we tested. The alloy wheel design is new for 1985.

build up revs, you run the risk of burning it out. First gear is suddenly very short, and you won't stay in any gear very long.

Callaway Engineering of Old Lyme, Connecticut has installed two IHI RHB5 turbos that push the charge through twin air-to-air intercoolers mounted on top of the engine. They are fed by a Callaway-designed air scoop, a well integrated fiberglass piece that distinguishes the conversion (along with the rear spoiler, "Callaway Twin Turbo" window lettering, BBS 16 x 7 wheels and Goodyear Eagle 205/55VR-16 tires). Callaway has also incorporated a solid-state fuel injection control it calls the Microfueler, which senses engine speed and manifold pressure to monitor the Bosch L-Jetronic injectors. All the stock Alfa emission controls remain. But the result is 230 bhp at 5500 rpm, a 50-percent increase over the stock engine, and 245 lb-ft torque at 2500. Overall the engine is very flexible and you can drive it virtually in 5th gear most of the time; cruising at 55 mph in top you're using 2500 rpm, just a bit less than the point where the fun starts. Even then you won't *have* to shift down to surprise those who don't know which Alfa this is.

We heard a little detonation at sustained full throttle, and when you back off there is noticeable turbo surge. On some turbos we've likened this sound to that of a respirator, but on the Callaway Alfa it's nothing less than the panting of a fire-breathing dragon.

Despite the Alfa's entertainingly responsive handling, several of our staff expressed more than a little respect for the available power, something they wouldn't want to learn about suddenly on an 80-mph curve. The chassis gives the driver the security of direct response. It can produce a diagonal pitching during fast cornering, but the steering is wonderfully precise and an educated use of the wheel and the throttle is what's called for. Nevertheless, we couldn't match the slalom speed of our previous GTV/6 test, the extra power being an excess in this case.

The ride is firm but acceptable; the Callaway's Goodyears give a good combination of ride and handling, although a true Big City pothole can give the whole structure a big bang.

To sum up the GTV/6, the stock version is a moderately improved product at a much more attractive price, appealing to the driving enthusiast but overdue for replacement by something more refined. (Sorry, all you *tifosi* out there, but it's true.) The Callaway Twin Turbo version will keep even the most skilled driving-with-a-capital-D enthusiast entertained for as long as his license lasts. It's the first Alfa in a long time (perhaps since the 8C 2900B, that supercharged master road car of the late Thirties) with the power to take on almost any rival. As equipped, the Callaway test car cost $27,234. That's half again as much as the standard GTV/6 but you do get half again the power, as well as full membership in the Supercar club.

Callaway Twin Turbo has the optional leather interior and boost gauge added below speedometer.

PRICE

	Callaway Twin Turbo	GTV 6/2.5
List price, all POE	$23,994	$16,500
Price as tested	$27,234	$17,395

Price as tested includes for the Callaway Twin Turbo, wheel and tire pkg ($2095), leather interior ($750), rear spoiler ($395); for the Alfa GTV 6/2.5, sunroof ($500), AM/FM stereo/cassette ($395)

GENERAL

Curb weight, lb	2890	2840
Test weight	3030	2955
Weight dist (with driver), f/r, %	51/49	50/50
Wheelbase, in.	94.5	
Track, front/rear	54.0/53.2	
Length	167.7	
Width	65.5	
Height	52.4	
Trunk space, cu ft	9.7	
Fuel capacity, U.S. gal.	17.6	

ENGINE

Type	turbocharged sohc V-6	sohc V-6
Bore x stroke, mm	88.0 x 68.3	
Displacement, cc	2492	
Compression ratio	7.5:1	9.0:1
Bhp @ rpm, SAE net	230 @ 5500	154 @ 5500
Torque @ rpm, lb-ft	245 @ 2500	155@ 3200
Fuel injection	Bosch L. Jetronic	
Fuel requirement	unleaded, 91/89 pump octane	

DRIVETRAIN

Transmission	5-sp manual	
Gear ratios: 5th (0.78)	3.20:1	
4th (0.95)	3.90:1	
3rd (1.26)	5.17:1	
2nd (1.96)	8.04:1	
1st (3.56)	14.35:1	
Final drive ratio	4.10:1	

CHASSIS & BODY

Layout	front engine/rear drive	
Body/frame	unit steel	
Brake system	10.4-in. (264-mm) discs front 9.8-in. (249-mm) discs rear; vacuum assisted	
Wheels	cast alloy, 16 x 7	15 x 6J
Tires	Goodyear Eagle 205/55VR-16	Pirelli P6 195/60HR-15
Steering type	rack & pinion	
Turns, lock-to-lock	3.5	

Suspension, front/rear: lower A-arms, upper lateral links with drag struts, torsion bars, tube shocks, anti-roll bar/De Dion axle on angled trailing links & Watt linkage, coil springs, tube shocks, anti-roll bar

CALCULATED DATA

Lb/bhp (test weight)	13.2	12.1
Bhp/liter	92.3	61.8
Mph/1000 rpm (5th gear)	21.8	
Engine revs/mi (60 mph)	2750	
R&T steering index	1.08	

ROAD TEST RESULTS

ACCELERATION

	Callaway Twin Turbo	GTV 6/ 2.5
Time to distance, sec:		
0–100 ft	3.0	3.4
0–500 ft	8.1	8.5
0–1320 ft (¼ mi)	14.5	16.4
Speed at end of ¼ mi, mph	97.5	85.0
Time to speed, sec:		
0–30 mph	2.1	2.6
0–50 mph	4.7	6.1
0–60 mph	6.2	8.5
0–70 mph	7.8	11.1
0–80 mph	9.7	14.0
0–100 mph	15.4	24.3

SPEEDS IN GEARS

5th gear (6300/5950 rpm) est	137	130
4th (6300)	115	
3rd (6300)	85	
2nd (6300)	55	
1st (6300)	30	

FUEL ECONOMY

Normal driving, mpg	18.5	25.0

BRAKES

Minimum stopping distances, ft:		
From 60 mph	157	153
From 80 mph	268	275
Control in panic stop	very good	
Pedal effort for 0.5g stop, lb	18	
Fade: percent increase in pedal effort to maintain 0.5g deceleration in 6 stops from 60 mph	22	
Overall brake rating	very good	

HANDLING

Lateral accel, 100-ft radius, g	0.81	0.75
Speed thru 700-ft slalom, mph	60.1	58.3

INTERIOR NOISE

Constant 30 mph, dBA	66	64
50 mph	69	66
70 mph	73	72

ACCELERATION

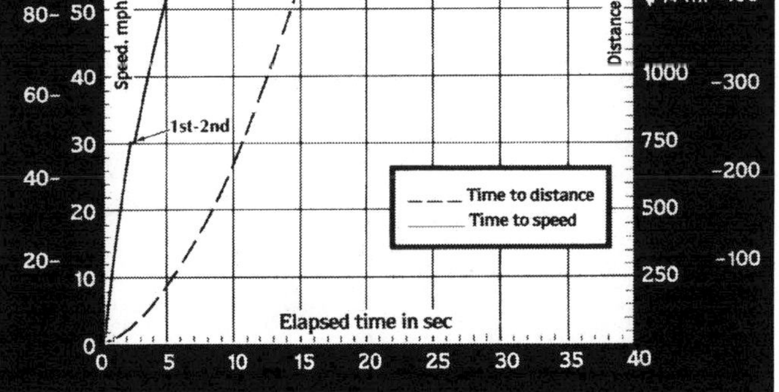

TRACK TEST

Bye bye Alfa

A nostalgic drive in one of Group A's most exciting racers, which is soon to be replaced by the Alfa 75

By Barry Lake

THE TWO YELLOW Alfa Romeo GTV6s of former world champion Alan Jones and long time star of local racing and rallying Colin Bond were the sensations of the 1985 Australian Touring Car Championship. Bond, the man in charge of preparing the works Alfas as well as being the regular team driver, and Jones put in some spectacular giant-killing drives. How could one ever forget AJ's spectacular run at Wanneroo in Western Australia in particular? It was nothing short of sensational. And both cars and drivers became noted for their dramatic but neat and controlled two-wheeling through corners. The yellow Ignis-sponsored cars were certainly good value for spectators and TV viewers alike.

Things have been tougher in 1986, with the bigger cars having gained more speed and with the BMW 325 of Tony Longhurst emerging as a front runner in the 2.5 litre class that was previously the Alfa's domain. The much faster Alfa 75 is anxiously awaited and will be here to replace this GTV6, hopefully, before the Sandown and Bathurst endurance races. In the meantime, Colin Bond and the GTV6 look like finishing their partnership in a blaze of glory following some stunning performances in the Better Brakes AMSCAR Series at Sydney's Amaroo Park.

"You realise, of course," said Alfa Romeo's PR supremo and race team manager Enrico Zanarini, "this is the fastest Group A Alfa Romeo in the world" and proceeded to hurl evidence at me of the car's better standing against the opposition in Australia than the same model enjoys in Europe.

"It's also the car that spent the least time in the pits at Bathurst last year", he went on, while pit manager John Gray and panel beater Keith Chesterton readied the car for my track test at Sydney's Oran Park circuit. "Our car made only two pit stops during the entire James Hardie 1000, for a tank full of fuel and four tyres each time, and was stopped for just over 30 seconds on each occasion. That's just over one minute in the pits," he continued.

The car we had at Oran Park this day was built new in Australia to replace the car written off by Bond in a crash at Surfers Paradise late last year. It was immaculately prepared, straight, tight as a drum, and Aussie-built except for as many parts as possible salvaged from the Italian-built Luigi car.

It usually runs on Pirelli tyres but for our test day it was mounted on Dunlop SP Sport radial slicks, 240 mm wide by 590 mm rim-to-tread height, mounted on nine-inch wide by 16-inch diameter wheels.

"You'll find they don't feel as positive as cross-ply slicks on initial turn-in," said Colin. "They tend to have a little 'walk' before they respond. Have you been driving any cars on slicks lately?" "No," I said, trying to remember the last time I'd been on slicks and unable to come up with a definite answer. "Oh, you'll find these great then," said Colin, who then gave a sharp burst of his inimitable laughter.

The car had just been trucked back from an ATCC round at Wanneroo in WA and had only been washed, had an old set of shock absorbers fitted in place of the slave units used for transporting, and had the tyre pressures checked.

The engine was as run at Wanneroo, last year's 187 kW unit that had to be put in for one race. A valve had dropped with disastrous results to the new 198 kW engine at Adelaide International Raceway the previous round. The extra power came from the allowance of unrestricted valve lift since January 1986. But 187 would be plenty for me on this test day.

Colin climbed into the car and

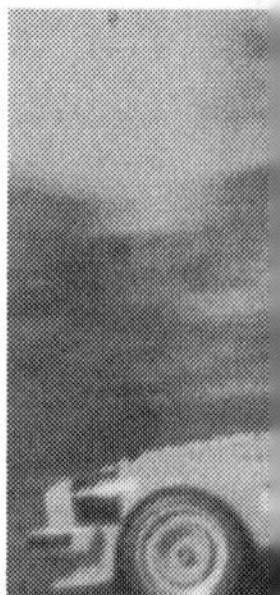

The editor gets a bit serious through the Oran Park Esses (above)

Photography by Kent Mears

whipped off half-a-dozen quick laps to check the car out. It was a general practise day just before an open meeting and the track was thick with Formula Vees, Formula Fords, Sports Sedans and Appendix J Touring Cars, but Colin got in some clean laps and his best was reeled off in 49.14 seconds.

So I was stuffed into the snug-fitting, flameproof Sparco seat and strapped in. "I think you'll find the driving position OK," said Colin. "It's too bad if you don't, we can't change anything." There was another burst of his raucous laugh.

Luckily, everything was a snug fit. The steering wheel in this car had been moved closer to the driver and the wheel-to-pedals ratio was much better for the average Aussie build than the road car's. It felt good.

The car was fitted with the standard tachometer, directly ahead of the driver. "Take it to 7500 revs," said Colin. "That's as far as the tacho goes anyway so just change gear when you get to the end of the dial." That's 1250 rpm past the standard red line but 500 rpm lower than Colin has used as a maximum in the past.

I knew I was going to be a bit busy to read all the rest of the gauges so I was glad to see two large amber warning lights in full view ahead that warn of dropping oil pressure or a lack of charge from the alternator. Further to the left was an equally large red light which operates when the turn indicators are flashing – to avoid accidentally confusing other drivers or the pit crew with an unintentional message should the flasher stalk be switched on during a hand-flaying slide through a corner.

Starting the car was a bit of a trick, too, and I had to learn the routine: pull the t-bar end of a cable to reconnect the battery isolator switch, then go to the multi-switch panel down on the transmission tunnel to flick on the ignition switch, press the starter button, and switch on the electric fuel pump . . . the latter has to be done last or the engine floods and won't start. That was worth knowing, as I discovered later. There are other switches on this panel to add to the confusion, one for a spare fuel pump, another for the demisting fan. On the dash there are also switches for such things as a transmission oil cooling system pump, but I was not going to need these in my short runs.

Colin assured me I wouldn't have any trouble warming up the slicks, but I had decided to hasten slowly anyway – there was no sense trying too hard too soon and getting ragged and unsafe.

But even on the standing start lap I felt immediately at home. This car felt so good, so well-balanced. It had 50/50 front to rear weight distribution with the 110-litre fuel tank full but we were running a low fuel load so there was a bit more weight bias towards the front.

On my second lap I gave it a bit of a squirt and I felt like I was well below the car's limit, so I gave it more on the third lap and then even more again on the fourth, fifth and sixth laps. I was just getting to the stage of sliding all four tyres across the road out of the bottom corner of Oran Park's short circuit, although I was deliberately staying below this level through the corners with concrete walls in close proximity. This was particularly costly in getting a slow entry to the main straight but I wasn't trying for a drive in a F1 team and my future as a track tester depended on re-

turning the car unscathed, so I was prepared to accept this.

I did feel, though, that I was braking too early at the end of the straight and, in the absence of anything solid to hit at this point, decided it was worth trying to go in deeper. The trouble was, I tried for too much in one hit. I was snaking around under brakes with the tyres just barely below the point of full lock-up, turned into the corner just a little late, the outside wheels were on the slippery, dirty outer section of the track, but I scrambled through with some opposite lock. But I was still off line and out of shape as I had to turn right into the Esses and the tail whipped out suddenly. I know the track well, we test there every week, and it's all smooth grass off this corner so I let the car have its way and speared off the track backwards, being careful not to lock up the wheels and flat spot the tyres.

I stopped unscathed, but had managed to stall the engine and had to run through the starting drill. Turn off the fuel pump, put the gear lever into neutral, press the starter button, the engine cranks over a couple of times and bursts into life, turn the fuel pump back on. Then select first gear, give it some revs, ease out the clutch and we're away again.

I did another lap to make sure everything was working, then a slowdown lap and stopped at the pits. "How silly do you have to be to spin a car that's as well balanced as this?" I asked Colin. "Oh, that's OK," he said. "Anyone can have a spin in a strange car. You probably could have held it, actually, 'coz the steering's really quick and you can get it really sideways and still catch it."

But I was off line and figured if I saved it there I might still spin further on where there are walls. I decided discretion was the better part of valour.

The car was needed at that stage to have some publicity photographs taken for Singapore Airlines so I had a chance to rest and think about the car. My son Gavin had been taking lap times. My first flying lap was 54.65 seconds, followed by 52.87, then 51.11, a 51.16, and 51.47 and then the spin. My best lap was my third flying lap and it was within two seconds of Colin's bogey time. I was very pleased with myself, so was Gavin. No one else had been taking any times or any notice . . . although Colin was impressed that I'd managed to restart the car. He thought for sure I'd forget to turn the pump off, flood it, and have to be rescued from my resting place.

After the photo session I got my chance for another drive. I felt more confident now and was immediately as fast as my previous best. My first three flying laps were 51.27 seconds, a 51.06 seconds, then my best at 50.74 and yet another 51 before I started to reel in an immaculate 1964 Appendix J Ford Mustang at the end of the straight. He

Examining the engine (above) and listening to Enrico Zanarini (below)

"And as I went backwards onto the grass . . . "

The editor behind the wheel (left) in third gear and heading for the straight. The Alfa's low gearing saw it through fourth and into fifth on the short straight. Third gear was used for all corners, with a brief snatch of fourth between The Esses and Yamaha Corner.

was braking really late and his lines through the corners were perfect, the car sliding – but not too much. By the Honda Dogleg, over the crest, I was right behind it and stuck there down the hill and through Castrol Corner. The Alfa was quicker and was being held up a bit, but I was impressed with this guy as he charged into the straight with all the V8 torque working to match the Alfa in acceleration. I figured I was going to be quicker by the end of the straight and should be able to brake later on this car's sticky slicks, and that's the way it worked out. But I was a bit concerned as I ranged up alongside him. Did he know I was there? Would he stay wide and give me room? How quickly could I enter the corner on this tight a line and survive?

Thankfully, the answers all came out

The Alfa's preparer and driver Colin Bond is sometimes quizzical (above), but more often laughing (below)

right and I was through. It wasn't until after we'd both stopped that I discovered it was Colin Bond driving the Mustang. He'd been asked if he'd like to drive a few laps in it – he would, and he did, and they rushed down to tell him they'd just clocked him at better than the Appendix J lap record around here . . . I had lost just over a second on that lap, but cranked off another couple of low 51s and then eased off to see where photographer Kent Mears was. I spent the next few laps concentrating on going fast only where he was taking pics, keeping out of other people's way, and wishing my arms would stop aching from the strain of steering this car with its high-geared steering (2.8 turns lock to lock). Colin says the steering's lighter – read "not as heavy" – with the cross-ply slicks on.

I was an even 1.6 seconds slower than Colin's time and felt I could get into 49s with more practice (resting my arms between runs) and if I booted it more quickly into the straight past that ever-so-close concrete wall, but Enrico was starting to make noises about packing up and we had some other cars to test and photograph for Modern MOTOR on the north circuit before the day was over, so it looked as though my time was up.

Colin saw me checking the lap times with Gavin, sauntered over to ask what had been my best lap, raised his eyebrows, asked what time he'd done earlier, then raised them a bit more and said: "I think I'll do a few more laps."

He reeled off a 49.26, then 48.01, a 48.41, a best of 47.91, a 48.28, then brought it into the pits. That put me in my place. Now I was 2.8 seconds off the pace and would have to be really working to get back to within two seconds of him again. But I wasn't going to get the chance. It was time to go.

At least I'd experienced the car's penchant for two-wheeling through the Dynamic Dip into Oran Park's Esses, comfortably clearing the inside kerb, not to mention this car's incredible balance, its neutral handling that sees both ends breaking away gradually when the tyres' adhesion level is reached – the rear just a little more than the front, pointing it into the corners.

It changes direction so well, steers and brakes well, despite needing lots of physical effort for both, the engine revs out freely to 7500 rpm, singing all the way in typical Alfa V6 style and has a good spread of torque that makes gear change points less critical than in many cars, the driving position is great, and the gearshift is far more positive than the standard car's, even though third gear slot wasn't as positively defined as I would like. The third gear synchro was slow, too, letting you know with a slight graunch when you didn't get the

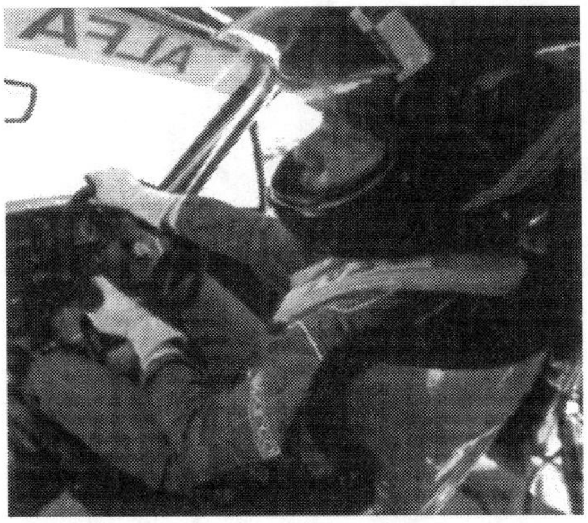

The Alfa needs little encouragement to lift two wheels (above and far left)

revs just right for the fifth-back-to-third change at the end of the main straight (only third, fourth and fifth gears were used here on the car's standard 4.1 to one final drive ratio).

These Group A cars are surprisingly little changed from standard, compared to full-race tourers of old, yet they're amazingly quick and well-behaved.

The engine of this Alfa has larger valves than standard, a modified camshaft for improved breathing, the compression ratio is up from 9.0 to one, to 11.0 to one, and the ports are cleaned up and matched to the manifolds within the homologated dimensions. Mahle pistons are fitted (the higher domes of which provide the compression ratio increase, they have cutaway clearances for the valves). The combustion chamber shape and volume remain standard. The bottom end is, of course, carefully assembled and balanced, and the cooling fan is removed to reduce unnecessary drag. An aluminium radiator is fitted for lighter weight and better cooling.

The rubber "doughnuts" in the tailshaft are replaced by universal joints for reliability; the brake discs are larger than standard – 279 mm at the front, 254 mm at the rear – and competition brake pads are used, in dual, single-spot calipers each side at the front and single twin-spot calipers per side at the rear; the brakes are operated by twin master cylinders with a brake bias bar between them – adjustable by the driver via a knob on the dashboard to vary the brake bias towards the front or the rear to avoid any tendency for one end to lock up before the other.

The suspension has cast aluminium lower control arms and larger (36 mm) torsion bars at the front; chrome-moly anti-sway bars front and rear; adjustable height coil springs for the rear de Dion tube which has rose-jointed ends for the trailing arms and Watt's linkage for more accurate location; and non-adjustable Bilstein gas-pressure shock absorbers are fitted all round.

The transmission has far closer ratios than the standard box, first being much higher and fifth much lower, at one to one, than the overdrive ratio fifth in the standard car. A limited slip differential is fitted and alternative final drive ratios are available; the standard 4.1 is used on most short circuits, but there is also a 3.81 and the 3.54 which is used at Bathurst and enables the car to reach its maximum speed of 241 km/h.

This is a fabulous little car that could make anyone feel good and though it will no longer be eligible for Group A racing after the end of this year, it would make a fantastic club racer – fast and reliable – and we hear the sale price was far less than the $70,000 the car cost to build initially.

The turbocharged Alfa 75 should take Bond into outright contention against the likes of Nissan's Skyline turbo and the Volvo turbo. □

ALFA ROMEO GTV6 GROUP A
2.5 litre, five-speed manual

ENGINE

Location	Front
Cylinders	60 degree V6
Bore x stroke	88.0 x 68.3 mm
Capacity	2492 cm³
Induction	Electronic fuel injection
Compression ratio	11.0 to 1
Fuel pump	Electric
Valve gear	Cog-belt driven single OHC per bank
Claimed power	187.0 kW at 7250 rpm
Claimed torque	253.0 Nm at 6500 rpm
Maximum recommended engine speed	7500 rpm
Specific power output	75.0 kW/litre

TRANSMISSION

Type	Five-speed manual
Driving wheels	Rear
Clutch	Twin dry plate

Gearbox ratios

Gear	Ratio	km/h 1000 rpm	Max Speed	At (rpm)
First	2.570	10.8	81	7500
Second	1.940	14.3	107	7500
Third	1.550	17.9	134	7500
Fourth	1.220	22.7	170	7500
Fifth	1.000	27.7	208	7500

Final-drive ratio	4.100 to 1
Top speed (Bathurst 3.54 to 1 final drive)	241 km/h

SUSPENSION

Front	Independent by upper and lower control arms with torsion bars and anti-roll bar
Rear	De Dion axle located by trailing arms and transverse Watts linkage with coil springs
Wheels	Alloy 9.0 x 16
Tyres	Dunlop SP Sport 240/590 R16

BRAKES

Front	265 mm discs
Rear	250 mm discs

STEERING

Type	Rack and pinion
Turns, lock to lock	3.6
Ratio	22.4 to 1
Turning circle	10.4 metres

DIMENSIONS AND WEIGHT

Wheelbase	2510 mm
Front track	1373 mm
Rear track	1352 mm
Overall length	4190 mm
Overall width	1650 mm
Overall height	1330 mm
Ground clearance	55 mm
Kerb weight	960 kg
Weight to power	5.1 kg/kW

CAPACITIES AND EQUIPMENT

Fuel tank	110.0 litres
Cooling system	12.0 litres
Engine oil system	6.6 litres
PRICE AS TESTED	$70,000

ALFA ROMEO GTV6 3.0

African Special

Only the red stripe hints at anything unusual about this GTV6.

John Dooley has a passion for Alfa Romeos. He drives them on the road for fun, and on the track as one of Alfa Romeo Dealer Team's top drivers in the RAC Saloon Car series. He also owns Brookside Garage which specialises in restoring older and particularly unusual examples of the marque, such as the GTA (the ultimate version of the Giulia GT racers), the SS with its exotic styling, and various Zagato-bodied types including the gorgeous, rapid, and desirable TZ. Having owned and driven such cars on road and track, it is not surprising that Dooley's everyday transport has to be an unusual Alfa — a 3-litre GTV6.

This is not a pre-production prototype of the forthcoming factory 3-litre, though; it is one of a run of 200 built in South Africa for homologation into that country's saloon car race series. With 2.5-litres, the standard GTV6 was being regularly beaten by BMW's 535i, so Alfa Romeo SA asked Milan what could be done. The answer was to take over a project which Autodelta (the now-defunct independent competitions department) had been developing. Since a 3-litre version of the V6 was up against punitive taxation on the home market, it had been shelved, but the company agreed to supply the necessary parts for AR SA to assemble.

Both bore and stroke were increased, so the special parts included cylinder head castings, crankshaft, pistons and sleeves, while larger valves were also fitted. The standard Bosch L-Jetronic fuel injection was replaced by the six carburetter set-up from the big Six saloon, and to clear the large air-cleaner, a new GRP bonnet was designed with an even bigger bulge in it than before, plus a central NACA duct just like on the Alfa Montreal. Ostensibly, this is to feed cold air to the carburetters,

ALFA ROMEO GTV6 3.0

Mid-range punch of this homologation special is phenomenal.

but on Dooley's car, they are fed from a normal pipe to the grille. Other differences are few: a slightly deeper front spoiler, lower ride

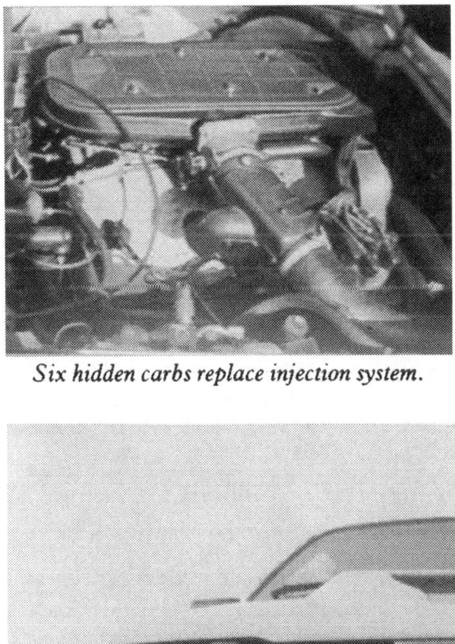

Six hidden carbs replace injection system.

height, some extra plastic body trim, and the discreetest of tail badges.

But successful though the cars proved on the track, it was not enough to stave off AR SA's economic problems, and last year the South African factory closed down. Most of the 3-litre cars had found ready homes in South Africa, but when Jon Dooley heard that there was one left he determined to get hold of it. As it was not Type-approved for the UK, it could only be privately imported as an effectively second-hand vehicle, so Jon made the long trip and ran the car in on its home ground, before freighting it to the UK.

I was keen to see the effect the extra 30 horsepower would have on the car that I have come to know so well with 160 bhp, and Jon readily obliged. It was odd to settle once again into that reclining crouch, knees spread and hands stretched to the wheel, but the long clonking gearshift was just the same. Not so the noise: the silky baritone chuckle is replaced by a more uncouth roar, laced with much obvious mechanical clatter. It feels rougher, too, right through the rev-range, but at the same time the urge of the 2.5 has become real punch. It seems more willing to spin its wheels than before, without actually taking off any faster (we recorded 0-60 mph times of 8.2 sec, almost identical to the standard car) but in third, fourth, and fifth the car surged ahead with the merest touch of the throttle. Intermediate times came out at 2.1 sec 30-50 mph, 3.3 sec for the next 20 mph increment, and 5.6 for the jump to 90 mph — very impressive indeed.

Converting this to real life, the car would swallow half-a-dozen vehicles at one gulp on the motorway and then find its driver stretching for sixth gear. On other roads, it appeared to be the mixture as before — sensitive steering which requires fairly large inputs, slow turn-in, and a deep desire to head for the opposite side of the road, though Jon was at pains to point out that strong braking and acceleration emphasise this. He is quite right, of course; the trouble is that the alternative, competition, approach of staying in a high gear and simply steering through the bend (which the car will do comfortably) requires absolute confidence that the tractor from the Ford advertisement is not going to mistake you for the Granada it usually pounces on.

It was a satisfying exercise to try this well-mannered special, particularly as it acted as a taster for the forthcoming factory alternatives of a 3-litre and a 2-litre V6. Although Jon Dooley's car is probably the last GTV6 to get a 3-litre unit, a 75 with a more refined version should be little short of sensational. **GC**

Huge bonnet bulge with NACA duct looks impressive.

ALFA ROMEO GTV COUPE

Alfa Romeo GTV6: *spare parts are reasonably priced, but service costs high. Later cars have much better build quality*

Although the addition of a body kit and Compotive wheels will not please the purists, the car is nevertheless eye-catching.

The Forgotten Sports Car

MIKE Kenyon of Chaparral Motorsport Ltd. of Westbury, Wiltshire has hit upon a very good idea. As the price of Ferraris seems to know no bounds and the gap between the exotic Italian sports cars and their more plentiful, but common, cousins widen, so certain models have increasingly appeared to be of good value. Although involved in other exotic makes of car, he increasingly became an Alfa Romeo specialist, with the result that, using the manufacturer's own criteria, he developed a defunct model into an even more thrilling sports car than many state of the art contemporaries.

The GTV in both 2-litre and V6 form is a model with which the staff at Standard House have long had a love affair with. Although none are to be found on the fleet nowadays, there was a time ten years ago when two were being run by various people.

It is a model which also has a history of attracting tuners as well. In July, 1980, we carried a report on the conversion to the 2-litre car carried out by West Horsley Alfa Romeo dealers Bell and Colvill. In conjunction with Mathwall Engineering, they offered a turbocharged version which boosted the power up to 175 bhp at 5500 rpm from the standard 122 bhp at 5400 rpm and increased the torque phenomenally with 190 lb.ft at 4000 rpm compared to 129 lb.ft at 4000 rpm. These figures even bettered Alfa's own Autodelta-developed turbocharged car which offered 150 bhp at 5500 rpm and 170 lb.ft at 4000 rpm.

In Australia Beninca Motors of Victoria offered a turbocharged version that was even more powerful, but the most exciting of all was the Twin Turbo GTV6 developed by Callaway Turbosystems of Connecticut, the same company that has developed the Aston Martin Virage engine as used in the Group C car as car be read elsewhere in this magazine. They transformed the car into a 230 bhp road burner which could slingshot to 60 mph in less than 6 seconds and hit 140 mph by the time the 6300 rpm redline was reached.

The GTV 2.5 was first announced ten years ago. At first, it seemed that the GTV coupé was being given no more than a fresh lease of life with the installation of the new V6 engine. The design, after all, was commenced in 1967 when Giorgetto Giugiaro, who had recently left Bertone and had started his own company Italdesign, and Aldo Mantovani, the body engineer were briefed by Alfa Romeo on their requirements. It was to be a coupé version of the saloon that was being developed by their own engineers.

The first designs were soon drawn up and development on the model, which had to go hand-in-hand with the saloon, continued through the final years of the Sixties and into the early Seventies. Apart from the design constraints imposed upon the designers by the necessity of using the same parts as the saloon, the car had also to be compatible in size to Fiat's successful 124 coupé with which the Milan company envisaged it was going to compete.

When Giugiaro's final designs and full scale models were accepted in early 1970, there was little that could be added apart from a few styling details. And yet Alfa Romeo's engineers wanted to have their own input, especially after a session in the wind tunnel with a prototype showed up what they thought were some deficiencies in the design. Although the ensuing modifications were relatively minor, Giugiaro was incensed enough to take legal action ensuring Alfa Romeo could not use his name in connection with the car.

The car's shape, now regarded as inspirational, did not receive universal praise at the time. Some felt that it was imbalanced and the wedge shape too pronounced - the windscreen too steeply raked and the bonnet too long. Even so, the car's aerodynamics were not particularly good, the Cd figure of 0.39 hardly awe-inspiring even by the standards of the day. Familiarity, however, tends to expunge from memory these earlier criticisms, especially when it is as subjective a matter as a car's looks, so that the shape is now held in generally high regard.

The Alfetta GT, as it was designated, was launched in May 1974, two years after the arrival of the Alfetta saloon. In the autumn of 1976, this 1779cc-engined version was superceded by two new models: the Alfetta GT 1.6, which utilised the 1570cc twin cam engine, and the top of the range Alfetta GTV, which used the 1962cc version of the aluminium twin overhead cam unit. These two models formed the basis of the model line-up for the next three years, the 1.6 lasting until 1979 while the 2000 GTV remained basically unchanged for the remainder of its life.

The next major development was the arrival of the GTV6, announced in late 1980 and launched in Britain in June 1981. It was the fastest production Alfa Romeo since the Montreal. Its main claim to fame was that it used an injected version of the V6 engine that had first seen the light of day in the Alfa 6. It was a model so loathed in Britain by the press, however, that few people were encouraged to sample the V6 engine, a unit that drew upon Alfa Romeo's racing experience. The bore, stroke and cylinder block design, for example, could trace their links to the racing department, as could the 46-degree valve angle which ensured better breathing. The original unit, however, was fed by 6 temperamental Dellorto carburettors which turned it into a moody beast.

The 60° cylinder placement, in place of the more conventional 90° configuration, ensured the engine could be fitted into the 4-cylinder compartment without too many alterations to the bodywork. The narrower vee also helped reduce torsional stress while exceptional balance was attained throughout the power range due to the equal distribution of all the combustion phases.

The Bosch L-Jetronic transformed the engine into a sweet-running, deliciously smooth unit and helped increase the power output to reach 160 bhp at 5600 rpm compared to 122 bhp at 5400 rpm of the Alfetta GTV. The torque figures were 157 lb. ft at 4000 rpm and 129 lb. ft at 4400 rpm respectively. It is the sheer artistry, the look, the noise and the heritage of this engine, however, which more than anything makes it such an exciting car, an obvious future "classic" if ever there was one.

Although Mike Kenyon is an enthusiast, he is a businessman first and foremost, and as with the wine connoisseur assessing future vintages, so Mike has come to the conclusion that the GTV6 is one of those forgotten sports cars whose time will surely come when they will be recognised for what they are.

The Alfa Romeos of the Seventies, made with the inferior Soviet steel, tend to be lost causes whereas the post '81 models, which were made of far superior European steel, are very much salvageable although they are still prone to rust and general decomposition. No matter the state of the donor car, therefore, once Chaparral lay their hands on it, it is completely stripped down to the bare metal so the corroded parts can be identified, for even the most outwardly perfect car can be hiding a veritable catalogue of disasters. Even with all their experience in dealing with these cars, examples still turn up bearing decay in previously unheard of places. The usual places, though, are the sills and inner and outer wings, while the bottom of the doors, the windscreen surround and the tailgate are closely inspected for decay. The crossmembers, however, once a real problem area, tend not to suffer.

No expense or effort is spared in replacing damaged or rotten panels with the result that by the end of this particular operation, the body shell is probably better than when it left the factory.

The paint shop is the next stop on the road to revival where it is etched with a special keying agent and given numerous coats of paint especially on the edges. All the known problem areas are especially treated with Hammerite paint to try and eliminate future trouble spots. The door uprights are powder coated, a process similar to stove enamelling involving the use of heat treatment, for added protection and wear.

While the shell is receiving all this treatment, the chassis and suspension are themselves being subjugated to minute inspection. The suspension parts are thoroughly cleaned and then powder coated and new rubber bushes are added. Adjustable gas shock absorbers are fitted which are set according to the customer's preference. Neither the De Dion type rear suspension, with coil springs, nor the twin wishbones at the front are usual problem areas, but the ball joint and bushes are replaced as necessary.

The engine is completely stripped to the bare block and fitted with new liners. New pistons, con-rods and the four main bearing, nitrided crankshaft are balanced and the cylinder heads are re-faced. The combustion chambers are all balanced and machined to give the right compression ratio. The ports are gas flowed as are the manifolds to match them while the valves are attended to. The manifolds are to original specification as there is not a great deal of room under the bonnet for any different sort.

The centre section of the exhaust is a standard Alfa Romeo system but the tail section manufactured by Amsa is used as it is believed to give a little more horsepower and it also gives a crisper note. All this work has helped boost engine output from 160 bhp at 6000 rpm to a stated 210 bhp.

The gearbox is stripped right down and any worn parts replaced before being re-assembled by an Alfa Romeo transmission specialist, the only modification made to the transmission being the installation of the single plate clutch from the current 3-litre car instead of the former dual-plate clutch.

Mike Kenyon believes that there is nothing intrinsically wrong with the much maligned transaxle gearbox, universally regarded as the Achilles heel of the car because of the lengthy linkages and the sometimes obstinate way it behaved. Great care is therefore taken on its reconstruction and installation so as to try and eradicate its faults.

With regard to the interior, Chaparral have a show car which is used as a basis, but the choice of material, colour and carpet and radio/cassette, which is not included in the price, completely remains the customer's choice.

When the car is stripped down, all the wiring is examined and connections checked. If there is any doubt as to its integrity, everything is replaced. At the suggestion that Italian electrics are not all that they might be, Mike Kenyon came back to his central theme that if the car is well put together in the first place, then most of the problems will be eliminated even on an Alfa Romeo, whether they are the gear linkage, rust or whatever. This also applies to the electrics. He asserts that many of the problems associated with the marque in the past are to do with poor quality control at the factory.

As far as the body is concerned, there was room for improvement. In trying to update the car visually, it was felt that as the original fixture of the bumpers was of poor design and execution, that here was an area that could greatly benefit from any modifications with the result that new front and rear bumpers, and side skirts, have been added. The rather unfortunate lump in the centre of the bonnet, which was necessary to clear the intake manifold, has been disguised with a special panel. The windscreens, door aperture rubbers and other such perishable items are replaced as a matter of course.

The original Campagnolo wheels have

The air intake manifold of the recently overhauled engine dominates the engine bay.

The twin exhaust pipe poking out from underneath the bumper and the bonnet bulge are the only signs that this is a V6-engined car.

been discarded for the British made, Compotive five spoke wheel shod not with the standard Pirelli P6 195/15s, but with 225/15 Goodyear NCTs which can accommodate 7" width tyres, a choice of rubber made following their racing experience with a GTV in the BARC Super Road Series.

The way Chaparral have gone about updating the car may not be to everybody's taste, body kits after all do tend to be frowned upon on everyday modern cars, let alone on "classics", but the styling suits the Alfa Romeo. There was no doubt that during the week I had the use of the car, it caused quite a stir wherever I went.

The test car was entirely refurbished in creamy white Connolly hide and black Wilton carpet. Although the seats were comfortable, memories came flooding back once in the driving seat. Less Italianate than remembered, the driving position still required the knees to be slightly splayed in order to get a comfortable grip on the steering wheel.

The pedals, though, were perfectly placed for heeling-and-toeing. Since this was a 1983 car, the dashboard was sensibly laid out with the larger tachometer and speedometer gauges flanking the oil pressure, water temperature and petrol tank gauges. Thankfully it was not one of the earlier versions which had the tachometer, later replaced by the speedometer, situated in splendid isolation in front of the driver while the rest of the instruments were located over the central console.

A two-speed wiper and wash/wipe cleared the screen but the car betrayed its early origins design-wise in its failure to have a wiper for the long, sloping rear hatch. A manual sun roof was fitted on this car which, although weather-proof, failed to keep the water out of the passenger side in a car wash, although fortunately I was alone at the time.

Although no more than a 2 plus 2, GTV6s were able to offer more rear legroom than many another sports coupé, but having said that, it has to be remembered that there is only sufficient room for passengers no older than young teenagers in the rear. One unfortunate aspect of the car, which is not in the remit of Auto-Technic, is the fact that the rear seats cannot fold down to create more space. The enlarged 16.5 gallon (75 litre) fuel tank of the GTV6 encroaches upon the area limiting boot capacity to just 7.9 cubic feet. Stowage space inside the car, however, is good.

The oversquare V6 never failed to burst into life at the turn of the ignition key, indeed the fuel injected unit was renowned for its good starting unlike the one in the Alfa 6 which had a reputation for being just the opposite. Even at tick-over, the burble from the powerful engine was music to the ears.

When cold, the oil pressure gauge needle flickered up to 175 lb. in on the extreme right of the dial before falling back to somewhere near the centre once the oil warmed up. To move off beforehand is to abuse the car and invite potential mechanical trouble.

Gingerly pushing the gear lever forward into first gear the expected graunch never materialised and it took some time to become acquainted with the fact that this was an Alfetta vice-free in this respect, even when selecting reverse. It was imperative, though, that the gearbox was allowed to warm up properly before putting too much load on it. Chaparral, however, are beginning to use AMSOIL in the gearbox, a synthetic oil which more quickly reaches an operating temperature. Obviously the use of such oils is clearly marked on the fluid reservoirs so that the incorrect oil is not used in future servicing.

At 3000 rpm in any gear the engine was muted, but the nearer it got to 4000 rpm, the more it began to show its true origin. Above 5000 rpm and the glorious howl of the V6, tiresome and loathesome to the *incognoscenti*, began to predominate so that conversation becomes a matter of sign language.

It is at this engine speed that the true nature of the V6 really shows itself. Although it has plenty of torque lower down the rev range, it is not until two thirds of the way up to rev band, which starts to be red lined at 5750 rpm, that maximum power comes on tap. Although the engine output is impressively high, Chaparral are more concerned with the feel of the car and its ability to get up to 100 mph. Driving along at the legal limit on the motorway, it was truly vibrant, ready to bound forward at the merest depression of the accelerator pedal.

We never had the opportunity to take the car onto a circuit or to Millbrook, but it is an established fact that the car when road tested when new had a top speed of 128 mph and could reach 0-60 mph in 8.2 secs. so there is every reason to believe that these figures can be easily exceeded. The most impressive factor, however, was its relative frugality. The overall mpg figure was just over 24 mpg and the touring mpg was 26.

The car was in its element in cross country and motorway driving. It was not really happy in town. The steering was not power assisted, but nevertheless felt perfectly weighted, the clutch was not too heavy and the engine was quite happy to burble around town, but it was like keeping a greyhound on a leash when taking it for a walk across Hampstead Heath. It was just begging to be worked hard.

Although the suspension can be adjusted to taste, that on the test car was set very low, perfect for rural roads, but not too good for ride comfort and urban situations. Even the relatively gentle sleeping policemen at Brent Cross Shopping Centre were a problem, the car scraping its underside no matter how carefully the obstacles were approached.

The handling was absolutely neutral. When cornering hard, tyre squeel would emanate from the front, but ultimately it would be the back that would break away. Straightline stability was good, only the higher than average tyre noise giving the game away that the car was shod with low profile Goodyear NCT tyres. At 3½ turns lock-to-lock, the steering is undergeared and required quite an amount of turn when approaching fast corners to counteract the initial understeer. Even in the wet, though, the grip was good

The brakes, disc all round, ventilated at the front and inboard at the rear, were light and responsive, the suspension set-up on the car never causing the nose to dip. Although brake pad wear should last 12,000 miles, GC found that when he had one of the early GTV6s on a long term test car in 1984, it tended to consume pads at a far quicker rate than that. Fortunately a situation never arose which needed emergency braking when driving the Auto-Technic car, but some heavy brake testing showed predictable handling and the brakes to be fade-free.

One problem with the car which Auto-Technic can do little about is that of condensation. The test car was taken over during a very wet spell. The heated rear window needed to be kept permamently on as well as a flow of warm air directed towards the windscreen by the two rather clumsy vent controls on the top of the dashboard. The two-position fan was noisy when in operation, but was necessary if the output from the heater was to be felt.

Although this car was a 1983 version, it was still a throwback to the Seventies. It was stylish, individual, powerful and Italian. It gave me so much fun that I was very sorry to return it at the end of the week. It had followed on the heels of a Porsche Carrera 2 that had been on test, and although as different as chalk is from cheese, it had given as much pleasure as the Stuttgart supercar.

The Alfa is very much a man's car - in looks, smell, noise, driving and character and could never hope to acquire the same universal appeal of some other, more mundane, sports cars. At present, the rebuild cost by Chaparral is £14,750, but a price that is very much dependable on the cost of the donor car.

It may sound expensive for a car that only went out of production in 1986 costing just over £13,000, but for that money you get a car of pedigree that has been re-built with great care and where the worksmanship is guaranteed for 12 months or 12,000 miles.

It may not be a "supercar", it may not even yet be a "classic" but it is the nearest thing to both without costing an arm and a leg.

WPK

Creamy Connolly hide with black piping and black Wilton carpet were the trim on the test car, but any material or colour can be specified.

Alfa daily

PHOTOS MATT BARNES

Alfa's smart coupés of the seventies and eighties make for stylish and rewarding classic motoring, and are surprisingly good value. Paul Clark looks at buying one

■ BUYER'S GUIDE: ALFA ROMEO GTV

Given a clean sheet of paper and asked to come up with a list of desirable features for their ideal Grand Tourer, what would the car enthusiast put down? Classic good looks? Perhaps. A thoroughbred engine? Maybe. Exhilarating performance? Probably. Charisma? Certainly.

Throw in fine handling, a five-speed 'box, cosy two-plus-two seating and a name that ranks with the very greatest in motoring history, and you have the essential ingredients that might persuade you to turn a blind eye to the inevitable faults associated with such a car.

That, in a nutshell, was what UK buyers of Alfa Romeo's new Alfetta GT got when they laid out over £4000 for the car when it first came to Britain in 1976.

Giugiaro's new hatchback design promised much in terms of enhanced practicality over the existing Bertone-styled coupé, while losing none of Alfa's renowned style. It wasn't a replacement for the Bertone car, which soldiered on until 1977, but it was clear that it would eventually supersede it.

Some features were controversial: a typically Italianate long arms/short legs driving position, not improved by the peculiar instrument layout, with the tachometer and supplementary gauges mounted in a central binnacle and the speedo directly ahead of the driver. Styling was clean and attractive, and the car boasted two-plus-two seating with a largish boot.

New seats were among features of 1981 styling change

Originally available only with the 1.8-litre 122bhp version of the famous twin-cam four, a 1.6-litre 109bhp model arrived in 1976, together with the 2000 GTV, fitted with the 1962cc engine and good for 130bhp.

A styling revision came in

Early GTVs had bizarre facia with centrally sited tachometer

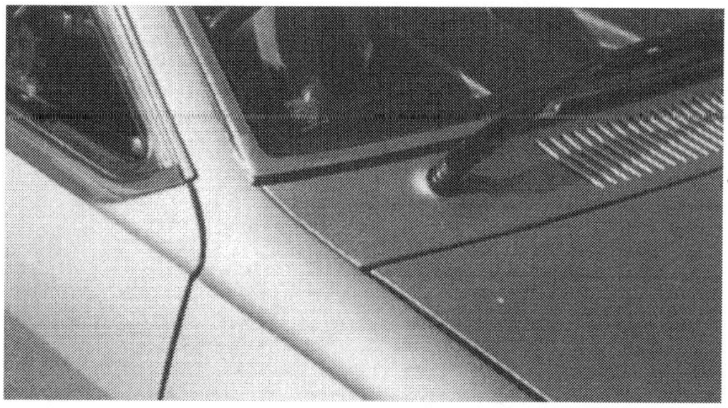

Cars tend to rust around 'screens and on top of front wings

GTV6s are most desirable, combining striking coupé looks with excellent performance

SPECIALISTS

EB Spares, 2 Washington Road, West Wilts Trading Estate, Westbury, Wilts BA13 4JP, tel 0373 823856.

Westune Service Ltd, 8 Howard St, Bolton, Lancs BL1 8HS, tel 0204 361571

Huntsworth, 24/28 Boston Place, London NW1, tel 071 724 0269

1981, with large and clumsy-looking plastic bumpers replacing the elegant stainless steel ones. New rear lights and a deep front spoiler were fitted, while a variety of plastic addenda, particularly on the rear quarter panels and beneath the rear bumper, broke up the purity of the original lines. Interiors were redesigned with a more conventional new facia and seats.

The big news for 1981, however, was the arrival of the GTV6, powered by the magnificent fuel-injected 2.5-litre V6 first seen in the ill-fated Alfa Six saloon.

Apart from the all-alloy V6 engine, which put out 158bhp, running gear was standard GTV except for a twin-plate clutch, larger, ventilated brake discs at the front and a bigger master cylinder.

Corrosion protection on these 'Series 2' cars was very much improved over earlier models and they continued in production, largely unchanged save for a further interior revision, until early 1987.

BODYWORK

It's true to say that any mid-seventies Alfa will be horrendously rust-prone, so you will be lucky to find an early car in really excellent condition unless it's been restored.

Particular rust points are around the front and rear screens, along the tops of the front wings, the rears and bases of the front wings, the tops and bottoms of the doors, the wheelarches and the trailing edge of the tailgate.

Look particularly for evidence of corrosion around after-market extras such as rear screen wipers, sunroofs (on earlier cars) and radio aerial apertures.

All body parts are common to both GTV and GTV6, with the exception of the GTV6 bonnet which had a bulge in it to accommodate the V6.

Replacement panels are available, most economically from David Edgington's EB Spares, noted spares specialists for a wide range of Alfas. Genuine front wings, until recently a problem, are now available at £170 each, sills at £38.50 a side, upper and lower front panels at £108 and £94 respectively, and rear wings (to special order only) at £240 each. A repair kit is available to repair damaged rear wheelarches at £18.50 each, with all these prices subject to VAT.

MECHANICALS

Although the GTV6 had ostensibly the same running gear as the GTV, there are numerous differences that make quite a difference to the cost of restoration and maintenance. Generally speaking, mechanical problems are much more expensive to rectify with the larger-engined car. That said, however, overall high quality

Left, early GTVs have 'pure' lines but rust away badly: it's very hard to find an excellent original car

Right, the GTV uses a 2-litre version of the famous Alfa twin-cam which produces a respectable 130bhp

means that reliability is good, so once a problem is properly solved, it is usually solved for good.

Starting with the engine, the main problems with the GTV6 are with the cylinder heads. Valve guides can wear prematurely, particularly on early cars (up to '84), leading to plenty of blue smoke on the overrun.

There's also a tendency for the multitude of small cylinder head oil seals to blow if the engine is revved hard when cold. Excessive oil pressure when cold is the culprit here, so always ensure the engine is fully warmed before revving above, say, 3000rpm.

Early cars were fitted with bronze valve guides and plain steel valves, later ones with steel guides and chromed valve stems. Although this modification cured the wear problem, the cost was considerably more – think in terms of up to £300 from EB Spares if you want to update your early spec engine to the later one. Alternatively, EB can supply original bronze guides for under £150 the set, which will last up to 50,000 miles.

A revised one-piece cylinder head gasket now replaces the original pair of gaskets, and this is said to eliminate the propensity for them to blow. On early cars check that gasket replacement has been carried out – it is almost routine maintenance at about 40-50,000 miles. It is as expensive as

Above, the V6 engine in the GTV6 was fuel injected

you might expect with a single overhead cam per bank V6, though obtaining the parts is no problem – a top gasket kit is £78 from EB Spares.

Properly maintained, the 2000 twin cam is a reliable and long-lived engine. A modest thirst for oil is nothing to worry about, and a rattle from the front may be nothing more than a slack timing chain.

Head gasket failure can be common on abused engines, so it is essential to ensure coolant levels and condition are maintained.

Remember that adjusting the valve clearances involves removing both cams, so check the engine is not too 'tappety' sounding, nor,

Right, high performance, yes, but there can be big bills too

indeed, suspiciously quiet, which may indicate the clearances are too tight. A slight tappet clatter is just right.

A set of standard pistons and liners for the 2000 will cost around £210, but EB carry a complete range of competitively priced uprated parts, including cams, carbs and manifolds, to enhance power output.

The GTV6 was fitted as standard with a twin-plate clutch which was notoriously troublesome and expensive. The plates had a tendency to stick on the splines, making engagement a very jerky affair. Too much slipping of the clutch by unsympathetic owners tended to burn out the first of the two plates, ruining the unit.

The standard modification nowadays is to fit the single-plate clutch from the 3-litre 75 saloon which improves both clutch feel and performance and is cheaper at around £220 complete with flywheel (twin-plate, £280). By contrast, the standard GTV single-plate clutch is about £85.

The gearshift is notoriously obstructive on the GTV, and Alfa never really cured the problem. Later cars

Chrome bumpers on GTV gave way to plastic on GTV6

Handling of all models is good, and can be made even better

Glenn Price

1977 GTV 2000

GLENN bought his GTV in 1979 and used it as daily transport until 1988 before taking it off the road for a full restoration. Since then the car has taken many concours awards at Alfa Romeo Owners Club events, including third overall at this year's National Alfa Day.

Undoubtedly the best early GTV in the country, the car is currently for sale at £7500-8000. Contact Glenn on 0664 69705 for further information.

David Marks

1985 GTV6

DAVID has owned his late-type GTV6 from new. The car is fitted with non-standard Compomotive wheels and 50-series tyres which, together with tuned suspension and Koni dampers, considerably enhance the roadholding and handling.

The engine has been fully blueprinted, and gas-flowed cylinder heads ensure that performance is much more impressive than standard while still maintaining perfect practicality.

(post-'84) are better than earlier ones in this respect but still nothing special.

It is important not to force the lever when the 'box is cold as damage can occur, particularly to first and reverse, which can jump out of gear, and second gear synchromesh. Standard driving practice when starting from cold is to move from first gear straight to third. A replacement bush set, price £7.50, is available from EB to rectify sloppiness at the base of the gearlever.

Brakes can be problematical, particularly the inboard rear calipers, which can suffer from inoperative adjusters or leakage past these adjusters. As no overhaul kits are available, replacement calipers are the only cure. David Edgington recommends new calipers at about £110 rather than secondhand ones which often retail at £80 or more. Be warned, though, that incorrect fitting can often ruin them.

Other points to watch for are an inoperative or erratic speedometer, often caused by a broken electronic sensor unit – an expensive item available only from Alfa dealers.

PRICES

Interest in these Alfas, particularly the GTV6, is mounting steadily and there's a ready demand for good examples.

You're unlikely to find a really good early GTV in original condition simply because so many have rusted to nothing. Average condition chrome bumper cars now change hands for anything between £200 and £1000 but expect to have to do a lot of work to keep it on the road. A good one should be no more than £2000-2500, and remember that these are probably faster than their later brothers.

Facelift models produced from late-1980 command more money because of their better build quality and rustproofing. Around £2000 should buy a sound example, and you can spend up to £3500 for an excellent one.

The GTV6 is currently enjoying a surge in interest, but are still excellent value as high performance GTs. Prices here start at around £2000 rising to around £8000 for one of the last (1987) cars in excellent condition with low mileage.

Post-1983 cars offer a higher specification, with revised seats and facia, a sunroof and electric door mirrors, and these command correspondingly higher prices.

With the potential high cost of repairs, it is sound practice to go for the best you can afford, rather than a 'dog' which needs plenty of work. Spend, say, £4000 on a sound example rather than £2000 on one needing work, and you will be better off in the long run. ■

Thanks to Alfa Romeo specialists Huntsworth, of 24/28 Boston Place, London NW1 (071 724 0269), for their help in the preparation of this article.

	GTV	GTV6
Engine	4-cyl, dohc all-alloy	60-degree V6, dohc all-alloy
Capacity	1962cc	2492cc
Bore/stroke	84 x 88.5mm	88 x 68.3mm
Max power	130bhp @ 5300rpm	160bhp @ 5600rpm
Max torque	129lb ft @ 4400rpm	157lb ft @ 4000rpm
Suspension	Front: ind by torsion bars, anti-roll bar, telescopic dampers Rear: De Dion tube, Watt linkage, coil springs	Front: ind by torsion bars, anti-roll bar, telescopic dampers Rear: De Dion tube, Watt linkage, coil springs
Brakes	Discs all round, servo-assisted	Discs all round, servo-assisted
Length	13ft 9in	13ft 9in
Width	5ft 5in	5ft 5in
Max speed	118mph	128mph
0-60mph	8.9sec	8.2sec
Fuel cons	25mpg	23mpg

BEHIND THE WHEEL

with Mick Walsh

Alfa GTV6

DESPITE THE FRUSTRATING foibles, the last two-and-a-half years years and 30,000 plus miles of GTV6 ownership have been a rewarding affair. Like most Alfas, it's a true driver's machine and on an early morning cross country dash it guarantees satisfaction. I still look forward to driving it which says a great deal for its addictive charm, never tiring of that musical rasp from the twin Ansa Italian tail pipes. Slide back the sun roof and that sweet V6 pitch is even clearer, a mechanical harmony that is at the heart of all great Alfas be it sohc V6, flat four or twin cam quartet. For that you will forgive the rest of the car almost anything.

I paid too much for C445SLH back in October '90 at the height of the 'investment' car boom and it's value has probably halved during ownership.

Read my contemporary road test of the GTV6 and misgivings about the notchy, sluggish gearchange are underlined. The transaxle unit and remote linkage is a concept Alfa have only recently mastered with the 75s. The GTV6 design isn't as bad as folk will have you believe although abused examples have now long since lost syncromesh in first and second. It's a change that requires patience particularly in the lower gears. Traffic light starts are handicapped unless you want to crash the cogs but out on the open road the engine's mid range punch more than compensates for the system's long travel. I enjoy double declutching which I'm sure has increased the longevity of my car's syncromesh and for the more experienced driver the pedals are perfect for heel and toeing.

The high pendulum pedal layout and particularly the high clutch action are another regular curse of the model. As with most Italian sportscars the steering wheel and pedals are too close for comfort. It's ideal for me with short legs and average arms, but tall figures are either too cramped in the footwell or arms uncomfortably stretched to reach the Momo leather wheel.

The clutch action is right at the peak of the pedal return which is regularly mistaken for a very worn unit. It's painful to sit with someone unused to the late take up particularly with twin plate fitment which doesn't tolerate clutch slip as windscreen stickers inform. I've recently exchanged the worn standard fitting with a single plate and although the clutch action is slightly heavier it does feel more positive. It's also stronger, and more important, around £70 cheaper. The official recommended tag for the twin plate is close to £600 so be warned.

The wedge bodywork still cuts it with todays soft look, but the interior, particularly the instruments look very dated with clumsy ergonomics. Early GTV6s copied its twin cam Alfetta sister with bizzare divorced speedo behind the wheel, but the revised facia reunited the dials in one cluster. As with all my Alfas, I tend to watch the oil pressure more than most gauges. The fuel early warning light was clearly devised for Italy and siesta gas station closure as it flashes annoyingly early.

But you never buy Alfas for practical reasons. The GTV6 is an inspiring car to drive. Performance is about equal with a 2.8i Capri with 0-60mph in around 7.6secs for which I suspect testers must have crashed the gears. With maximum torque at 4000rpm and power (160bhp) at 6000rpm its at the top end the GTV6 gives most pleasure. The seductive mid-range punch allows for relaxed fast motoring, and matched to impressive chassis balance and assured grip it adores smooth open sweepers. Hustle it impatiently and the front end understeers in protest but a smooth pilot can balance out the handicap with the throttle.

Soft brake pedal pressure and snatching anchors (possibly over served) are another odd balance that takes adjustment but the vented discs (inboard at the rear) have always performed impressively. My only other misgiving, apart from the unsightly advances of rust worm, was an unnerving float at high speed particularly on motorways but a conversion to Spax shocks all round has transformed that uneasy sensation without impairing the ride.

Today you can buy the very best GTV6s for around £6000 complete with all the desired performance tweaks, and given the chance a second time around that's the car I would go for. The recent clutch replacement and suspensions mods pushed my running costs to nearly £3000 over two-and-a-half years. Now I'm nervous of expensive valve guide work around the corner not to mention bubbling steel.

Alfa's driving position very Italiante, but suits Walsh who never tires of driving the car

Printed in Dunstable, United Kingdom

91370215R00058